AN IN-DEPTH GUIDE TO MOBILE DEVICE FORENSICS

AN IN-DEPTH GUIDE TO MOBILE DEVICE FORENSICS

Chuck Easttom

CRC Press is an imprint of the
Taylor & Francis Group, an **informa** business

First Edition published 2022
by CRC Press
6000 Broken Sound Parkway NW, Suite 300, Boca Raton, FL 33487-2742

and by CRC Press
2 Park Square, Milton Park, Abingdon, Oxon, OX14 4RN

ISBN: 978-0-367-63298-4 (hbk)
ISBN: 978-0-367-63300-4 (pbk)
ISBN: 978-1-003-11871-8 (ebk)

DOI: 10.1201/9781003118718

Typeset in Times
by MPS Limited, Dehradun

Contents

Section I Technical Details of Mobile Devices

Section II Forensic Techniques

Section III Additional Topics

Section I

Technical Details of Mobile Devices

1 Wireless Basics

Wireless communications are fundamental to cellular technology. That includes the essentialls of radio wave technology as well as cellullar networks. Cell networks including GSM, EDGE, UMTS, LTE, and 5G. There are other wireless technologies included in mobile phones. These technologies include 802.11 Wi-Fi as well as Bluetooth.

INTRODUCTION

It should be relatively obvious that cell phones use wireless communications. However, we must be more specific in order to understand cellular technology. It is important to define the type of wireless communication in question. That will require us to examine how the electromagnetic waves themselves behave. Mobile forensics often focuses on just the device. And we will certainly be doing that in later chapters. However, it is important that you also understand the actual signals being transmitted. This chapter will provide a rigorous introduction to these topics so that you can better understand the cellular technology introduced in the rest of this book.

ELECTROMAGNETIC WAVES

Wireless communication of any type requires using electromagnetic waves to transmit the signal. Our first goal will be to understand the electromagnetic spectrum, then we will delve into how to encode data in such transmissions. It is important to have a fundamental understanding of the basic physics of electromagnetic transmissions in order to understand precisely how mobile devices send and receive data.

Electromagnetic waves also known as electromagnetic radiation (EMR) are the waves of an electromagnetic field propagating through space. Such waves include visible light, radio waves, infrared and more. The complete spectrum of electromagnetic radiation is shown in Figure 1.1.

Notice that as the figure moves from left to right, the wavelength gets shorter, and thus the frequency gets longer. This is a fundamental fact about electromagnetic waves. They all travel at the same speed, the speed of light, which is approximately 300,000 km/second or 186,000 miles/second. Since the speed is constant, changing frequency or wavelength must change the other measurement proportionally. In Figure 1.2 you can see both wavelength and frequency illustrated.

At the top you see a wave with a large wavelength. Note that both waves have the same total length. The wave on the top only has four peaks in this space of time, whereas the wave on the bottom has eleven peaks. Clearly the wave on the bottom has a higher frequency and shorter wavelength. It should also be noted that higher

DOI: 10.1201/9781003118718-1

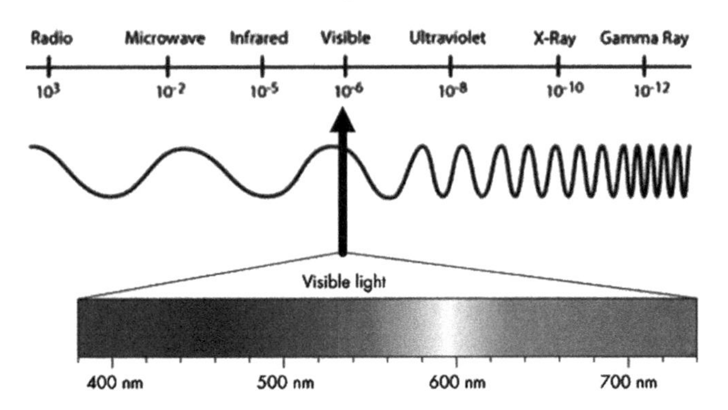

FIGURE 1.1 The electromagnetic spectrum.

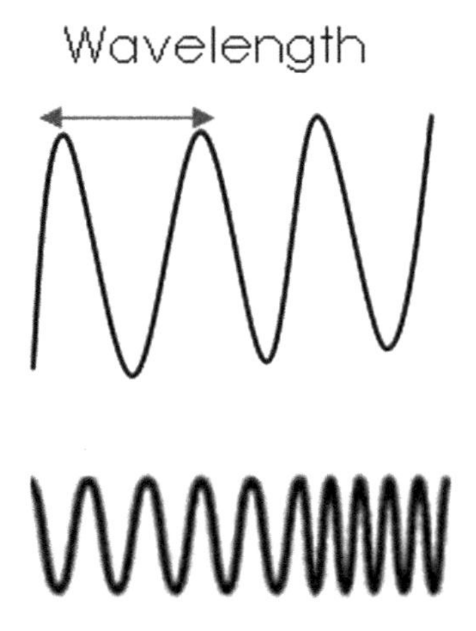

FIGURE 1.2 Frequency and wavelength.

frequency means more energy. Consider Figure 1.1 again. With wavelengths on the order of 10^3 meters we have radio waves, which are not harmful. You encounter radio waves all the time. However, on the far right we have X-Rays with a wavelength of 10^{-8} meters and gamma rays with a wavelength of 10^{-12} meters. These are both quite harmful. The relationship of frequency and wavelength is defined by this simple equation shown in (1.1):

$$f\lambda = c \tag{1.1}$$

f is frequency
λ is wavelength
c is the speed of light

Radio transmission, as well as other types of electromagnetic radiation communication, works on the principle of producing a carrier wave as the means of communication between two wireless devices. The *carrier wave* is an electromagnetic wave of a particular frequency that is used to carry data. Technically you could use any frequency to encode and transmit data. However, for rather obvious reasons gamma rays and x-rays make a poor choice for communications. Frequency assignments are something you are likely to be familiar with. Radio and television broadcast stations are assigned specific frequencies on which they must transmit. A radio identification such as 101.1 FM represents a carrier wave of 101.1 MHz (megahertz). The frequency of 101.1 MHz also happens to be the frequency for the classical music station in my area.

It is important that you fully understand the hertz unit of measurement. A hertz is the international standard unit for frequency and is defined as one cycle per second. A cycle is moving from one peak to the next (or alternatively from one trough to the next) in a wave. A more formal definition is the number of oscillations per second. As you can see from the previous discussion of radio communication, we are normally discussing thousands or millions of cycles per second. Thus, we will usually talk about kilohertz (kHz), megahertz (MHz), gigahertz (GHz), or even terahertz (THz). Frequency is not only used for electromagnetic radiation, but also for sound waves. For example, an adult human, of normal hearing capacity, can detect sound between 20 Hz and 16,000 Hz. When you hear terms such as ultrasound, that literally means frequencies beyond normal human hearing (ultra). The American National Standards Institute has actually defined ultrasound as a bit higher than that, sound frequencies greater than 20 kHz or 20,000 Hz.

The carrier wave is encoded with data. This is done through some type of *modulation.* The technique of modulation is how AM radio, FM radio, and all electromagnetic communication works. In face AM means amplitude modulation, and FM means frequency modulation. A simple radio broadcast consists of a *transmitter,* which generates the carrier wave and modulates information into the carrier wave, and a *receiver,* which receives the modulated wave and demodulates it. The transceiver and receiver must both be at or very near the same carrier wave frequency for communication to occur. The receiver decodes the data that was encoded in the carrier wave. This process is known as *demodulation.*

The same principle is used to transmit digital data signals. In fact, you probably used such signals today. Not only is this how your cellular phone operates, but it is also how your wireless internet works. A carrier wave establishes the transmitter/receiver relationship. The carrier wave is modulated with a wave pattern resembling the digital data signal. The two waves are combined before transmission and then separated at the receiver.

RADIO WAVES AND COMPUTER NETWORKING

Radio waves are electromagnetic waves with a frequency range of 10 kHz to 3,000,000 MHz Radio waves are used in wireless networks. Radio wave-based networks adhere to the IEEE 802.11 standards and Bluetooth standards and operate at 2.4 GHz.

To communicate between radio-based wireless network devices, several transmission techniques are used: single-frequency, spread-spectrum, and orthogonal frequency-division multiplexing. These, or similar techniques, are used with encoding any data into electromagnetic radiation. The techniques are based on the technology, the frequency band of operation, and the manufacturer's idea of the best way to achieve a high data rate. A high data rate not only relies on how fast the data can move between two points, but also on how much data has to be retransmitted because of interference.

Spread spectrum is a transmission method that uses multiple channels to transmit data either concurrently or sequentially. The term *spread spectrum* refers to transmission channels *spread* across the *spectrum* of the available bandwidth. The spread spectrum method that transmits data on multiple channels simultaneously is called *frequency hopping*. Frequency hopping is more technically called frequency hopping spread spectrum (FHSS). This works by using an allocated band of frequencies and hopping between the frequencies, using one frequency at a time. The signal transmitted rapidly changes the frequencies among the available frequencies in the bandwidth. This is referred to as hopping.

Frequency hopping is used with wireless devices that use the 2.4-GHz radio band. The 2.4-GHz frequency has a bandwidth of 83.5 MHz. Instead of utilizing the entire range as a single channel to carry radio data, the frequency band is divided into seventy-nine 1-MHz channels. Rather than transmitting the data packets over a single channel, the data packets hop from one channel to another in a set pattern determined by an algorithm. None of the channels are used continuously for more than 0.4 seconds. Since the data packets switch from frequency to frequency, this transmission technique is called *frequency hopping*.

The spread spectrum method that transmits data on multiple channels sequentially is called *direct sequencing*. This is more technically referred to as direct sequence spread spectrum (DSSS). This technique uses a pseudorandom bit sequence to differentiate the signal from random noise at the same frequency. Direct sequencing divides the 2.4-GHz frequency band into eleven overlapping channels of 83 MHz each. Within the eleven channels are three channels with a 22-MHz bandwidth. The three channels do not overlap and can be used simultaneously. Using three channels simultaneously results in higher data rates than are achieved with frequency hopping. The data rates for direct sequencing are 11 Mbps and 33 Mbps. The 33 Mbps is a result of using all three 22-Mbps channels at the same time. One disadvantage of direct sequencing is that a much larger portion of the transmitted data is affected by electromagnetic interference than with frequency hopping. The data rate of direct sequencing, therefore, is substantially impacted by interference.

While FHSS and DSSS are among the most well-known methods of encoding information, there are others. One example is the *chirp spread spectrum* method (CSS). CSS uses the entire bandwidth to broadcast a signal, as do all spread spectrum methods. CSS relies on a sinusoidal signal of frequency increase or decrease, called a chirp. For readers not familiar, sinusoidal is another way of saying the wave is a sine wave.

Time hopping spread spectrum (THSS) is another method. With this method, a pseudorandom number sequence is used to vary the period and cycle of the carrier wave. This method is less common than the other methods we have discussed. This technique is used for anti-jamming. It is also difficult to intercept.

Spread spectrum is the preferred transmission method of most wireless technologies. There are variations of spread spectrum, but all share the common principles just discussed. Transmitting data on multiple channels decreases the likelihood of interference. Interference is typically limited to only one or two of the channels. The other channels in the frequency band are free to carry data undisturbed. Data that is lost can be easily retransmitted on a channel that is not affected by the interference.

The *orthogonal frequency-division multiplexing (OFDM)* transmission method is utilized with wireless devices that operate in the 5-GHz radio band. The OFDM transmission approach divides the allotted frequency into channels. This is comparable to the division done in frequency hopping and direct sequencing. *Orthogonal*, used in this context, means there are multiple distinct radio channels side by side within a designated radio band. *Frequency division* means to divide the allocated frequency range into multiple, sub-frequencies. *Multiplexing* means to combine content from different sources and then transmit them together over a single carrier. By combining the three methods into one single process, OFDM is designed to communicate wireless data over several different channels within an assigned frequency range. However, in OFDM, each channel is broadcast separately and is referred to as *multiplexed*.

OFDM was first introduced by Bell Labs in 1966. It has been used in DSL internet, power line networks, and both 4G and 5G networks. That final item, the 4G and 5G networks, makes ODFM particularly relevant for our discussions. OFDM is used in combination with the Unlicensed National Information Infrastructure (U-NII) frequency ranges. The FCC divided the 5-GHz radio frequency into three channels of 20-MHz. These were classified as the Unlicensed National Information Infrastructure (U-NII). The three classifications are U-NII-1, U-NII-2, and U-NII-3. Each of the three U-NII classifications has a frequency range of 100 MHz. Using the OFDM transmission technique, each 100-MHz frequency range is broken down into four separate 20-MHz channels. Each of the 20-MHz channels is further divided into fifty-two, 300-kHz sub-channels. Forty-eight of the fifty-two sub-channels are utilized to transmit data. The remaining four sub channels are used for error correction. The large number of channels provide the high data rates.

The FCC U-NII classifications are based on a few requirements. These requirements include the frequency range of the broadcast as well the allowable maximum amount of power allotted to the broadcast. Another issue impacting classifications is the location of where the device may be used. There is no maximum distance measurement in feet or meters for the different classifications. The maximum distances are controlled, to a large extent, by the maximum amount of output wattage that can be generated by the devices. The actual range varies considerably due to items such as building structures and materials, the electromagnetic environment, and atmospheric conditions.

The maximum range achieved in transmission will vary by vendor and based on location conditions. Location conditions include environmental issues such as mountains and large buildings. The maximum power yield of the device has a direct

TABLE 1.1
Communication Frequencies

Source	Frequency
Super Low Frequency (SLF)	30–300 Hz
Ultra-Low Frequency (ULF)	300–3,000 Hz
Very Low Frequency (VLF)	3–30 kHz
Amateur Radio	1.8–29.7 MHz
Citizens band (CB)	27.065 MHz to 27.405 MHz (40 channels)
Amateur Radio	28–30 MHz
Land mobile	29–54 MHz
Amateur	50–54 MHz
TV low VHF	54–88 MHz
Land mobile (EU)	65–85 MHz
FM BCB (J)	76–90 MHz
FM BCB (US & EU)	88–108 MHz
Aircraft	108–136 MHz
TV high VHF	174–216 MHz
TV UHF	470–806 MHz
Cellular AMPS	806–947 MHz
Amateur Land mobile GPS	1200–1600 MHz

relationship to data throughput. This will also affect range. At least some packet loss is due to radio interference or excessive distance between two devices. When packet loss increases, the data rate decreases.

There is a built-in process for adjusting to help prevent packet loss. The data rate is automatically adjusted to a lower rate when an excessive number of packets are lost. The data rate continues to be lowered until an acceptable packet loss is reached. The more powerful the signal, the more resistant it is to interference. Thus, there will be fewer packets lost. Therefore, data rate is better when the signal is stronger. The maximum transmission power rating for a wireless device is set by the FCC.

In Table 1.1 you can see many of the common frequencies used for various types of communication. As you can see there is a wide array of communication frequencies. Each of these is used for a different purpose.

There is a much wide range of frequencies for cellular communication. Some of the common LTE frequencies are shown in Table 1.2.

As can be seen in Table 1.2, the frequency range varies based on the specific technology and the location. You can probably surmise that 5G has a similar range of available frequencies. Table 1.3 describes some of the more common frequencies used in 5G.

TABLE 1.2
LTE Frequencies

Band	Common Name	Uplink (MHz)	Downlink (MHz)	Duplex Spacing (MHz)	Channel Bandwidths (MHz)
1	IMT	1920–1980	2110–2170	190	5, 10, 15, 20
2	PCS (Personal Communications Service)	1850–1910	1930–1990	80	1.4, 3, 5, 10, 15, 20
3	DCS (Digital Cellular System)	1710–1785	1805–1880	95	1.4, 3, 5, 10, 15, 20
4	AWS-1 (Advanced Wireless Services)	1710–1755	2110–2155	400	1.4, 3, 5, 10, 15, 20
5	Cellular (CLR)	824–849	869–894	45	1.4, 3, 5, 10
7	IMT-E	2500–2570	2620–2690	120	5, 10, 15, 20
8	Extended GSM	880–915	925–960	45	1.4, 3, 5, 10
11	Lower PDC (Personal Digital Cellular)	1427.9–1447.9	1475.9–1495.9	48	5, 10
12	Lower SMH (Seven Hundred Megahertz)	698–716	728–746	30	1.4, 3, 5, 10
13	Upper SMH	777–787	746–756	−31	5, 10
14	Upper SMH	788–798	758–768	−30	5, 10
17	Lower SMH	704–716	734–746	30	5, 10
18	Lower 800 (Japan)	815–830	860–875	45	5, 10, 15
19	Upper 800 (Japan)	830–845	875–890	45	5, 10, 15
20	Digital Dividend (EU)	832–862	791–821	−41	5, 10, 15, 20
21	Upper PDC	1447.9–1462.9	1495.9–1510.9	48	5, 10, 15
24	Upper L-Band (US)	1626.5–1660.5	1525–1559	−101.5	5, 10
25	Extended PCS	1850–1915	1930–1995	80	1.4, 3, 5, 10, 15, 20
26	Extended Cellular (CLR)	814–849	859–894	45	1.4, 3, 5, 10, 15
28	APT (Asia Pacific Telecom)	703–748	758–803	55	3, 5, 10, 15, 20
29	Lower SMH	N/A	717–728	N/A	3, 5, 10
30	WCS (Wireless Communication Service)	2305–2315	2350–2360	45	5, 10
31	NMT (Nordic Mobile Telephone)	452.5–457.5	462.5–467.5	10	1.4, 3, 5
32	L-Band (EU)	N/A	1452–1496	N/A	5, 10, 15, 20
34	IMT (International Mobile Telecom)	2010–2025		N/A	5, 10, 15
35	PCS	1850–1910		N/A	1.4, 3, 5, 10, 15, 20
36	PCS	1930–1990		N/A	1.4, 3, 5, 10, 15, 20

(Continued)

TABLE 1.2 *(Continued)*
LTE Frequencies

Band	Common Name	Uplink (MHz)	Downlink (MHz)	Duplex Spacing (MHz)	Channel Bandwidths (MHz)
37	PCS	1910–1930		N/A	5, 10, 15, 20
38	IMT-E	2570–2620		N/A	5, 10, 15, 20
39	DCS–IMT Gap	1880–1920		N/A	5, 10, 15, 20
40	S-Band	2300–2400		N/A	5, 10, 15, 20
41	BRS (US)	2496–2690		N/A	5, 10, 15, 20
42	CBRS (EU, Japan) (Citizens Band Radio Service)	3400–3600		N/A	5, 10, 15, 20
43	C-Band	3600–3800		N/A	5, 10, 15, 20
44	APT	703–803		N/A	3, 5, 10, 15, 20
45	L-Band	1447–1467		N/A	5, 10, 15, 20
46	U-NII-1–4	5150–5925		N/A	10, 20
47	U-NII-4	5855–5925		N/A	10, 20
48	CBRS (US)	3550–3700		N/A	5, 10, 15, 20

MULTIPLE ACCESS

In addition to frequency issues, there are issues with multiple access. Essentially more than one phone wants to access the cell tower at a given time. The base stations need to serve many mobile terminals at the same time (both downlink and uplink). That alone would cause a multiple access issue. At the same time, all mobiles in the cell need to transmit to the base station. There is also the issue of interference among different senders and receivers. What is needed is a multiple access scheme. There are three methods of doing this. Some of the names will sound similar to how spread spectrum frequencies are handled. The three multiple access methods are called:

- Frequency Division Multiple Access (FDMA)
- Time Division Multiple Access (TDMA)
- Code Division Multiple Access (CDMA)
- Quadrature Division Multiple Access (QDMA)

While the names are similar to what is used for spread spectrum, the purpose and implementation is quite different. Let us first look at Frequency Division Multiple Access. Each mobile unit is assigned a separate frequency channel for the duration of the transmissions. Often a mobile device will use two channels, one for uplink and one for downlink. There is a guard band between frequencies to prevent interference from adjacent frequencies. Guard bands are simply unused radio

TABLE 1.3
5G1 Frequencies

Band	Uplink (MHz)	Downlink (MHz)	Duplex Spacing (MHz)	Channel Bandwidths (MHz)
n1	1920–1980	2110–2170	190	5, 10, 15, 20, 25, 30, 40, 50
n2	1850–1910	1930–1990	80	5, 10, 15, 20
n3	1710–1785	1805–1880	95	5, 10, 15, 20, 25, 30, 40
n5	824–849	869–894	45	5, 10, 15, 20
n7	2500–2570	2620–2690	120	5, 10, 15, 20, 25, 30, 40, 50
n8	880–915	925–960	45	5, 10, 15, 20
n12	699–716	729–746	30	5, 10, 15
n13	777–787	746–756	–31	5, 10
n14	788–798	758–768	–30	5, 10
n18	815–830	860–875	45	5, 10, 15
n20	832–862	791–821	–41	5, 10, 15, 20
n25	1850–1915	1930–1995	80	5, 10, 15, 20, 25, 30, 40
n26	814–849	859–894	45	5, 10, 15, 20
n28	703–748	758–803	55	5, 10, 15, 20, 30
n29	N/A	717–728	N/A	5, 10
n30	2305–2315	2350–2360	45	5, 10
n34	2010–2025		N/A	5, 10, 15
n38	2570–2620		N/A	5, 10, 15, 20, 25, 30, 40
n39	1880–1920		N/A	5, 10, 15, 20, 25, 30, 40
n40	2300–2400		N/A	5, 10, 15, 20, 25, 30, 40, 50, 60, 70, 80
n41	2496–2690		N/A	10, 15, 20, 30, 40, 50, 60, 80, 90, 100
n46	5150–5925		N/A	10, 20, 40, 60, 80
n47	5855–5925		N/A	10, 20, 30, 40
n48	3550–3700		N/A	5, 10, 15, 20, 40, 50, 60, 80, 90, 100
n50	1432–1517		N/A	5, 10, 15, 20, 30, 40, 50, 60, 80
n51	1427–1432		N/A	5
n53	2483.5–2495		N/A	5, 10
n65	1920–2010	2110–2200	190	5, 10, 15, 20, 50
n66	1710–1780	2110–2200[5]	400	5, 10, 15, 20, 25, 30, 40
n70	1695–1710	1995–2020	300	5, 10, 15, 20[B 4], 25[B 4]
n71	663–698	617–652	–46	5, 10, 15, 20
n74	1427–1470	1475–1518	48	5, 10, 15, 20
n75	N/A	1432–1517	N/A	5, 10, 15, 20, 25, 30, 40, 50
n76	N/A	1427–1432	N/A	5
n77	3300–4200		N/A	10, 15, 20, 25, 30, 40, 50, 60, 70, 80, 90, 100

(Continued)

TABLE 1.3 ***(Continued)***
5G1 Frequencies

Band	Uplink (MHz)	Downlink (MHz)	Duplex Spacing (MHz)	Channel Bandwidths (MHz)
n78	3300–3800		N/A	10, 15, 20, 25, 30, 40, 50, 60, 70, 80, 90, 100
n79	4400–5000		N/A	40, 50, 60, 80, 100
n80	1710–1785	N/A	N/A	5, 10, 15, 20, 25, 30, 40
n81	880–915	N/A	N/A	5, 10, 15, 20
n82	832–862	N/A	N/A	5, 10, 15, 20
n83	703–748	N/A	N/A	5, 10, 15, 20, 30
n84	1920–1980	N/A	N/A	5, 10, 15, 20, 25, 30, 40, 50
n86	1710–1780	N/A	N/A	5, 10, 15, 20, 40
n89	824–849	N/A	N/A	5, 10, 15, 20
n90	2496–2690		N/A	10, 15, 20, 30, 40, 50, 60, 80, 90, 100
n91	832–862	1427–1432	570–595	5, 10
n92	832–862	1432–1517	600–660	5, 10, 15, 20
n93	880–915	1427–1432	527–547	5, 10
n94	880–915	1432–1517	532–632	5, 10, 15, 20
n95	2010–2025	N/A	N/A	5, 10, 15
n96	5925–7125		N/A	20, 40, 60, 80
n97	2300–2400	N/A	N/A	5, 10, 15, 20, 25, 30, 40, 50, 60, 80
n98	1880–1920	N/A	N/A	5, 10, 15, 20, 25, 30, 40

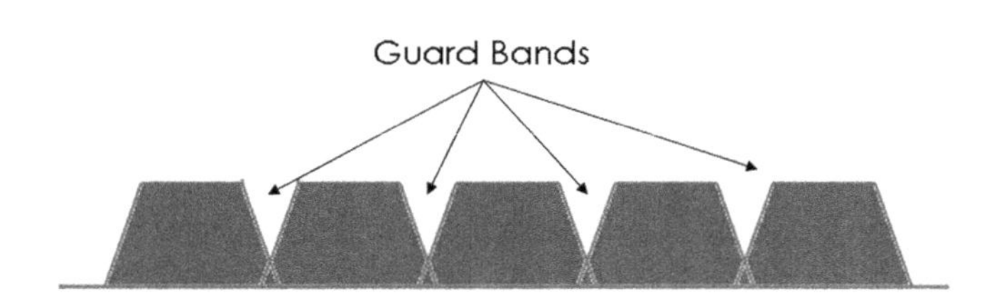

FIGURE 1.3 FDMA guard band.

spectrum separating the frequencies that are being used. Figure 1.3 illustrates the guard band used in FDMA.

TDMA is another method that is used to share the available bandwidth. Time is divided into slots and only one mobile terminal transmits during each slot. Each user is given a specific slot. No competition in cellular network. Essentially each device is given a very small sliver of time, and all devices are rotated through. The time slots are small enough that no user notices that they don't have all the time slots. Just as there are guard bands in FDMA, there are very small-time segments that are guard time slots in TDMA.

TDMA is used in 2G and GSM networks. It is also often used with PDC (Personal Digital Cellular). Time is first divided into frames, and the frames are further divided into slots. Frames are measured in milliseconds. Slot times are frequently measured in microseconds. DTDMA or Dynamic Time Division Multiple Access is a variation which uses a scheduling algorithm to reserve the time slots in each frame based on traffic demand. DTDMA is used in Bluetooth, WiMAX (IEEE 802.16a) and in some military radios.

Yet another method is Code Division Multiple Access (CDMA). Orthogonal codes are used to separate different transmissions. Each symbol to be transmitted is encoded using a specific code. Since each user is using a different code, they can all share the bandwidth. As you can probably surmise, the choice of what code to use is rather critical. The best performance is found when all the users have substantial differences in their signal. That is directly impacted by the selection of codes.

Quadrature Division Multiple Access is not as widely known, it is used primarily in short range applications. This method combines quadrature phase shift keying with CDMA. Phase shift keying (PSK) conveys data by modulating the phase of the carrier wave. Bluetooth and Radio Frequency ID (RFID) both use PSK. For those readers not familiar with phases, let us take a moment to describe them. Phases are a more complete description of a transmission. For two transmissions to be in the same phase, they would need to have the same wavelength, same cycle, and same frequency. Thus, phase shifting, is shifting to a different phase. Figure 1.4 illustrates two waves that are out of phase by 90 degrees.

QAM (Quadrature Amplitude Modulation) is often used in video systems but can be used to encode any digital data into an analog medium. It conveys two analog message signals, or two digital bit streams, by changing (modulating) the amplitudes of two carrier waves, using the amplitude-shift keying (ASK) digital modulation scheme or amplitude modulation (AM) analog modulation scheme. QAM is actually a family of modulation methods. It works by modulating the amplitudes of two carrier waves using either amplitude shift keying or amplitude modulation. The two waves are out of phase to each other by 90 degrees. Mathematically, the term for 90 degrees out of phase is orthogonal.

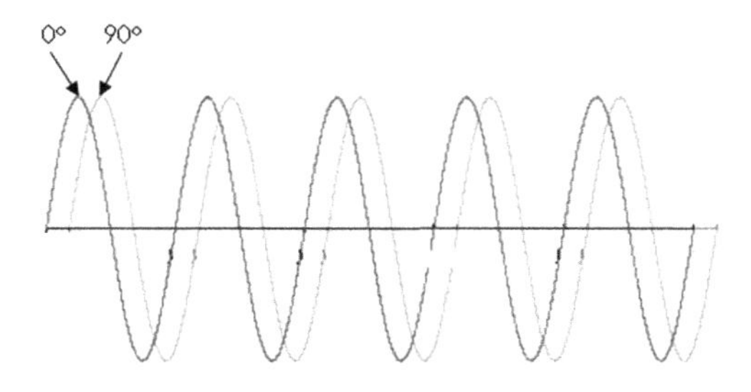

FIGURE 1.4 Out of phase.

QAM uses a constellation diagram. This is a representation of the signal being modulated. It is often used to represent QAM of phase shift keying. The signal is displayed in a two-dimensional scatter diagram. The number of symbols on the diagram indicates the number of symbols that can be transmitted with each sample.

The constellation diagram provides a great deal of information. One half the distance between each pair of neighboring points is the amplitude of noise or distortion that would cause one of the points to be misidentified as the other. This would cause a signal error. This means that the further the points are separated from one another the more resistant they are to noise. The following figure is a 4-QAM constellation. An example of a constellation diagram is shown in Figure 1.5.

One issue to keep in mind is co-channel interference. Multiple channel transmissions can have co-channel transmissions. The formula for co-channel interference is shown in Equation 1.2.

$$\frac{S}{I} = \frac{S}{\sum_{k=1}^{N_i} \langle I_k \rangle} \tag{1.2}$$

In Equation 1.2 there are several symbols that need to be defined:

S = desired signal power in a cell (note that many texts use "C" instead of S).

I_k = interference signal power from the kth cell.

N_i = number of interfering cells.

The signal power at any point is inversely proportional to the inverse of the distance from the source raised to the Υ power. $(2 < \Upsilon < 5)$

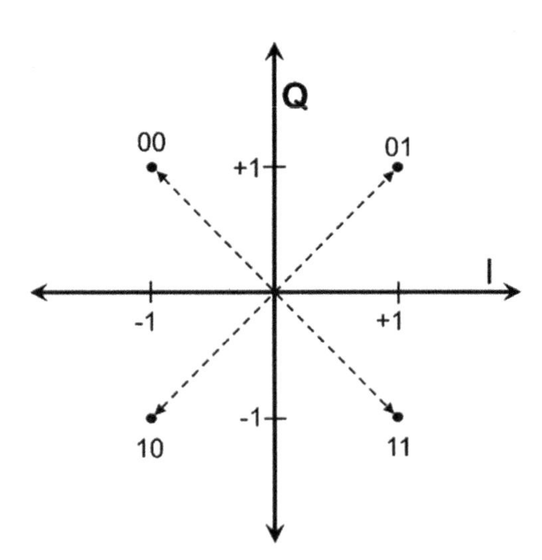

FIGURE 1.5 QAM.

CELLULAR TECHNOLOGY

Cellular technology uses radio waves, the electromagnetic radiation we have been discussing earlier in this chapter. Coverage area is divided into designated areas referred to as *cells*. Phones do not communicate directly with each other. Rather each phone communicates with a tower that provides connection between sender and receiver. Space is divided into cells, and each cell has a base station (tower and radio equipment). Base stations transmit to and receive from mobiles at the assigned spectrum. If you move from one cell to another, the first cell notices your signal strength decreasing, the second cell notices your signal strength increasing, and they coordinate handover so your handset switches to the latter cells

There are actually five primary types of cells:

- Macro-cell: their coverage is large and high-power transmitters and receivers are used
- Micro-cell: their coverage is small and are used in urban zones; low-powered transmitters and receivers are used to avoid interference with cells in other clusters
- Pico-cell: covers areas such as building or a tunnel
- Femto-cell: covers a home or small business, or something of similar size.
- Umbrella-cell: Many smaller cells are grouped under a larger cell, the umbrella cell.

Micro and macro cells are the most common. However, picocells are becoming more common. They are useful in providing cellular communication in areas that might not otherwise be reachable. A classic example is a long tunnel.

CELLULAR NETWORKS

Cellular network technology has evolved. 5G is the latest version of cellular technology. However, you will find older networks still in operation. LTE or 4G is still quite common. And in some areas you may encounter 3G or even 2G. The various networking types are summarized in the following subsections.

GSM

Global System for Mobile communications. This is a standard developed by the European Telecommunications Standards Institute (ETSI). Basically GSM is the 2G network. GSM was first deployed in 1991 in Finland. GSM supports five different cell sizes: macro, micro, pico, femto, and umbrella cells. Cell tower radius depends to some extent on antennae height. The largest practical coverage area is approximate 22 miles/35 kilometers.

GSM utilized a range of frequency bands. However, regardless of frequency TDMA is used for access to the GSM network. Frames are approximately 4.615 ms and are divided into eight channels with each channel having a rate of approximately 270.8 kbits/second.

EDGE

Enhanced Data Rates for GSM Evolution. This one does not fit neatly into the 2G/3G/4G spectrum. It is technically considered pre-3G but was an improvement on GSM (2G). Therefore, one could consider it a bridge between 2G and 3G technology. Edge uses a different type of encoding called Gaussian Minimum Shift Keying (GMSK) as well as PSK/8 Phase Shift Keying. Gaussian minimum shift keying works similarly to minimum shift keying. The digital data is first put through a Gaussian filter then the frequency is modulated. This tends to give narrower phase shift angles. High order PSK is yet another variation of phase shift keying. This sometimes allows for alternative modulations such as the quadrature amplitude modulation discussed earlier in this chapter.

GSM security is defined in the 3GPP standards. The 3GPP standards will be covered later in this chapter. However, for now we can discuss the security of GSM. The cryptographic algorithms used with GSM are: A5/1, A5/2, and A5/3. These are stream ciphers used in cellular communications. All GSM phones support A5/1 and A5/2, however, A5/1 is more common in networks.

A5/1 is used in the United States and Europe. It was first developed in 1987, before GSM became widely used. GSM transmissions are organized in bursts. Each channel typically sends on burst every 4.615 seconds, and each burst has 114 bits of information. A5/1 is used to combine the 114-bit sequence with a keystream of 114 bits. The combination is done using the binary exclusive or (XOR) operation. For those readers not familiar with the XOR operation, a brief discussion of binary operations is provided. XOR is important as it is used in many different cryptographic algorithms, including those in cellular phones:

The binary AND operation combines two binary numbers, one bit at a time. It asks is there a 1 in the same place in both numbers. If there is, then the resulting number is a 1. If not, then the resulting number is a 0. You can see in Figure 1.6a results of a binary AND. Starting from the far right, the least significant bit, the top and bottom numbers are combined using the binary and operation.

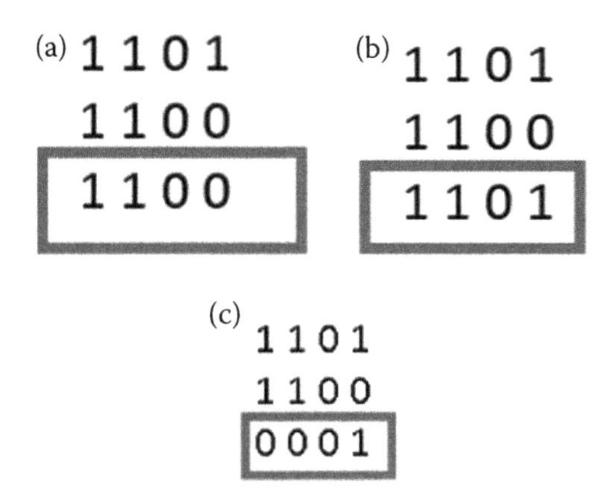

FIGURE 1.6 (a) Binary AND, (b) Binary OR, (c) Binary XOR.

The binary OR operation is similar with a slight difference. It is asking the question of whether there is a 1 in the first number or the second number, or even in both numbers. So basically, if either number has a 1, then the result is a 1. The result will only be a zero if both numbers are a zero. You can see this in Figure 1.6b.

The binary XOR is the one that really matters for cryptography. This is the operation that provides us with interesting results we can use as part of cryptographic algorithms. With the binary exclusive or (usually denoted XOR) we ask the question is there a 1 in the first number or second number, but not in both. In other words, we are exclusively asking the "OR" question, rather than asking the "AND/OR" question that the OR operation asks.

The XOR operation is used in cryptography because it is reversible. If you XOR the result (0001) with the second number (1100) you get back the first number (1101).

Now that we have covered the binary operations, we can return to the A5/1 algorithm. A5/1 uses three linear feedback shift registers (LFSR). An LFSR is a register (hardware component of a processor). These registers are used for several purposes, including generating pseudo random numbers. For A5/1 it is the pseudorandom number function that is important. That allows the generation of a random key for the XOR operation with the data in the burst.

UMTS

Universal Mobile Telecommunications Systems. This is a 3G standard based on GSM. It is essentially an improvement of GSM. UMTS uses a variation of code division multiple access called wideband code division multiple access (W-CDMA). The W-CDMA transmits on a pair of radio channels that are each 5 MHz wide. W-CDMA was first deployed in Japan in 2001. UMTS theoretically supports data transfer rates of up to 42 Mbits/s. However, this is not what users typically experience. A term you will see frequently in reference to UMTS is UTRAN (UMTS Terrestrial Radio Access Network). This is a term for the network and the equipment that connects the mobile devices to the public switched telephone network and the internet.

UMTS has standards for mutual authentication and key agreement. UMTS uses a smart card application called the Universal Subscriber Identity Module (USM) working with the network to authenticate the subscriber to the network, and the network to the subscriber. Authentication begins with a pre-shared secret key that is stored in the Authentication Center (AuC) of the network, and in the USIM of the device. The pre-shared key will be used to help create a cipher and integrity key. The cipher and integrity key are new for each session.

More specifically, the AuC and the USIM share some specific components:

- The secret key specific to each subscriber
- Authentication algorithms that consist of authentication functions (f1, f1*, and f2)
- Key generation functions (f3, f4, f5, f5*).

The AuC has a random number generator and a sequence number generator. The USIM has an algorithm to assure that sequence numbers received from the AuC are new. The specific parameters used in UMTS authentication are given here:

- K = Subscriber authentication key (128 bit)
- RAND = User authentication challenge (128 bit)
- SQN = Sequence number (48 bit)
- AMF = Authentication management field (16 bit)
- MAC= $f1_K$ (SQN||RAND||AMF) (64 bit)
- (X)RES = $f2_K$ (RAND) = (Expected) user response (32–128 bit)

- CK = $f3_K$ (RAND) = Cipher key (128 bit)
- IK = $f4_K$ (RAND) = Integrity key (128 bit)
- AK = $f5_K$ (RAND) = Anonymity key (48 bit)
- AUTN= SQN⊕AK||AMF||MAC (128 bit)

The ⊕ symbol denotes the exclusive or (XOR) process discussed earlier in this chapter.

The primary cryptographic algorithm used is the UMTS Encryption Algorithm (UEA1). This algorithm makes use of the KASUMI block cipher. KASUMI was specifically designed for the 3GPP standard. The full UMTS 3GPP specification can be found at https://www.etsi.org/deliver/etsi_ts/133100_133199/133102/05.00.00_60/ts_133102v050000p.pdf

LTE

Long Term Evolution. This is a standard for wireless communication of high-speed data for mobile devices. This is what is commonly called 4G. LTE is implemented with different bands and frequencies in different countries. Therefore, to use the same phone in all countries, the phone must be a multi-band phone.

There are variations on the basic LTE technology. Long-Term Evolution Time-Division Duplex (LTE-TDD) is one such technology. LTE-TDD uses a single frequency and alternates uploading and downloading. The ratio of upload to download can be changed dynamically based on current conditions. The frequency band for LTE-TDD is shown in Table 1.4.

Another approach Long-Term Evolution Frequency-Division Duplex (LTE-FDD). LTE-FDD utilizes paired frequencies to upload and download data. LTE-FDD. LTE Frequencies are shown in Table 1.5.

LTE has security improvements over UMTS. Some of the characteristics of LTE Security are given here:

- Re-use of UMTS Authentication and Key Agreement (AKA)
- Use of USIM required (GSM SIM excluded, but Rel-99 USIM is sufficient)
- Extended key hierarchy
- Possibility for longer keys
- Greater protection for backhaul
- Integrated interworking security for legacy and non-3GPP networks

TABLE 1.4
LTE-TDD Frequencies

Band No.	Frequency Band	Band Width
33	1900 MHz–1920 MHz	20 MHz
34	2010 MHz–2025 MHz	15 MHz
35	1850 MHz–1910 MHz	60 MHz
36	1930 MHz–1990 MHz	60 MHz
37	1910 MHz–1930 MHz	20 MHz
38	2570 MHz–2620 MHz	50 MHz
39	1880 MHz–1920 MHz	40 MHz
40	2300 MHz–2400 MHz	100 MHz
41	2496 MHz–2690 MHz	194 MHz
42	3400 MHz–3600 MHz	200 MHz
43	3600 MHz–3800 MHz	200 MHz
44	703 MHz–803 MHz	100 MHz
46	5150 MHz–5925 MHz	775 MHz

Before delving into the key exchange process, a few acronyms must be defined. MME is Mobility Management Entity. This is the key control node for the LTE network. HSS is the Home Subscriber Server. HHS is a central database that contains user and subscription information. This is the LTE version of the HLR. SGSN is the Serving GPRS Support Node. GPRS is an acronym for General Packet Radio Service. GERAN is an acronym for GMS EDGE Radio Access Network. Now with these acronyms in mind, the chart shown in Figure 1.7 should make some sense to you.

5G

5th-Generation Wireless Systems (abbreviated 5G) Meets ITU IMT-2020 requirements and 3GPP Release 15 Peak Data Rate 20 Gbit/second Expected User Data Rate 100 Mbit/second. Speeds have ranged from around 50 Mbit/second to over a gigabit/second. Due to the increased bandwidth, it is expected that 5G networks will not just serve cellphones like existing cellular networks, but also be used as general internet service providers, competing with existing ISPs such as cable internet and provide connection for IoT. 5G NR (New Radio) is a new air interface for 5G and is the global standard for air interfaces for 5G network. A summary of 5G NR cells and environments is given in Table 1.6.

CELL SYSTEM COMPONENTS

As was seen in the previous section, cellular technology has evolved. However, the basic components for cellular systems are more or less the same. In this section we will summarize those components.

TABLE 1.5
LTE-TDD Frequencies

Band No.	Uplink Frequency Band	Downlink Frequency Band
1	1920 MHz–1980 MHz	2110 MHz–2170 MHz
2	1850 MHz–1910 MHz	1930 MHz–1990 MHz
3	1710 MHz–1785 MHz	1805 MHz–1880 MHz
4	1710 MHz–1755 MHz	2110 MHz–2155 MHz
5	824 MHz–849 MHz	869 MHz–894 MHz
6	830 MHz–840 MHz	875 MHz–885 MHz
7	2500 MHz–2570 MHz	2620 MHz–2690 MHz
8	880 MHz–915 MHz	925 MHz–960 MHz
9	1749.9 MHz–1784.9 MHz	1844.9 MHz–1879.9 MHz
10	1710 MHz–1770 MHz	2110 MHz–2170 MHz
11	1427.9 MHz–1447.9 MHz	1475.9 MHz–1495.9 MHz
12	699 MHz–716 MHz	729 MHz–746 MHz
13	777 MHz–787 MHz	746 MHz–756 MHz
14	788 MHz–798 MHz	758 MHz–768 MHz
15	Reserved	Reserved
16	Reserved	Reserved
17	704 MHz–716 MHz	734 MHz–746 MHz
18	815 MHz–830 MHz	860 MHz–875 MHz
19	830 MHz–845 MHz	875 MHz–890 MHz
20	832 MHz–862 MHz	791 MHz–821 MHz
21	1447.9 MHz–1462.9 MHz	1495.9 MHz–1510.9 MHz
22	3410 MHz–3490 MHz	3510 MHz–3590 MHz
23	2000 MHz–2020 MHz	2180 MHz–2200 MHz
24	1626.5 MHz–1660.5 MHz	1525 MHz–1559 MHz
25	1850 MHz–1915 MHz	1930 MHz–1995 MHz
26	814 MHz–849 MHz	859 MHz–894 MHz
27	807 MHz–824 MHz	852 MHz–869 MHz
28	703 MHz–748 MHz	758 MHz–803 MHz
29	N/A	717 MHz–728 MHz

Mobile Switching Center (MSC) This is the switching system for the cell network. These can be used in 3G or in GSM networks (you will learn about both of those terms later in this section). The MSC processes all the connections from both mobile devices and land line calls. It is also responsible for routing calls between base stations and the public switched telephone network (PSTN).

Base Transceiver Station (BTS) This is the part of the cell network responsible for communications between the mobile phone and the network switching system. It consists of a base transceiver station and a base station controller. Some sources

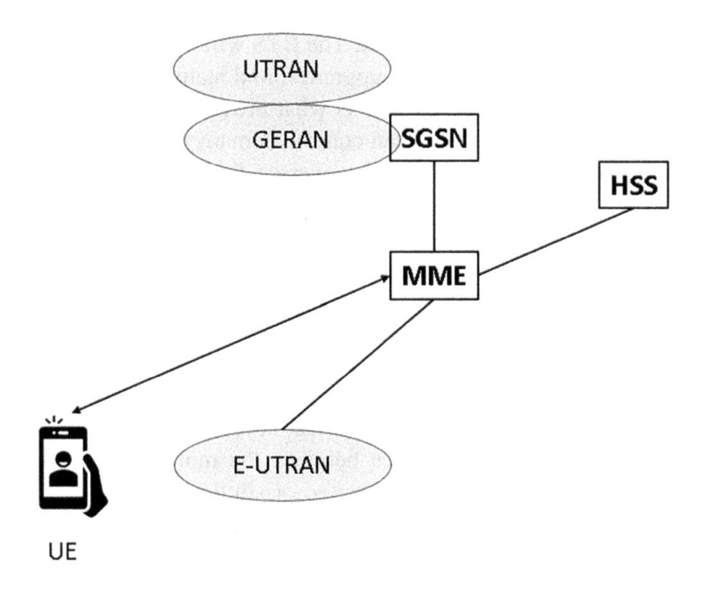

FIGURE 1.7 LTE authentication.

TABLE 1.6
5G NR Cells

Cell Types	Deployment Environment	Max. Number of Users	Max. Distance from Base Station
Femtocell	Homes, businesses	Home: 4–32	tens of meters
Pico cell	Public areas such as airports or shopping malls	64–128	tens of meters
Micro cell	Urban areas to fill coverage gaps	128–256	few hundreds of meters
Metro cell	Urban areas to provide additional capacity	more than 250	hundreds of meters

refer to the BTS as the base station (BS). In 3G networks the BTS is sometimes called the node B. The BTS is one component of the Base Station System. The BSS Base Station System is radio transceiver equipment that communicates with cellular devices. The BSC is a central controller coordinating the other pieces of the BSS. The BTS is controlled by a base station controller using the base station control function (BCF).

A BTS will have several components. The most obvious is the transceiver. This is often coupled with a power amplifier that amplifies the transceiver signal. Another obvious component is the antenna. There is also a combiner that combines the feeds from several transceivers. Multiplexers are responsible for sending and receiving signals to and from the antenna. The BTS will also have a control function that manages things such as software upgrades, and status changes.

Base Station Controller (BSC) This is what provides the thinking behind Bast Transceiver Stations. A single BSC can control as many as several hundred BTS's. One important function of the BSC is to oversee the handover from one BTS to another BTS. Among other things, the BSC contains a database that has information on all carrier frequencies, frequency hoping lists, and other information critical to mobile communications.

Base Station Subsystem (BSS) This is the core of the mobile network handing signaling and traffic between cell phones and towers. It has a number of components such as the BTS and BSC that we have already discussed. The cells can be sectorized by simply using directional antennae at the BSS, with each antennae pointing in different directions. The BSS has several interfaces. An example is the Um interface. This is the interface between the mobile station and the BTS. Mobile station can be a cell phone, computer, or similar device. The A interface is between the BSC and MSC and carries traffic channels. The Abis interface connects the BTS and BAS. There are other interfaces, but these should give you a general idea of the interfaces used in the BSS.

Home Location Register (HLR) The database used by the MSC for subscriber data and service information. It is related to the Visitor Location Register (VLR) which is used for roaming phones. In LTE HLR was replaced with HHS Home Subscriber Server.

Subscriber Identity Module (SIM) This is a circuit that stores the International Mobile Subscriber Identity (IMSI). Think of it as how you identify the phone. Many modern phones have removable SIM, which means you could change out the SIM and essentially have a different phone with a different number. A SIM card contains its unique serial number (ICCID), the IMSI, security authentication and ciphering information. This SIM will also usually have network information, services the user has access to and two passwords. Those passwords are the personal identification number (PIN) and the personal unblocking code (PUK).

Personal Unlock Number (PUK) This is a code used to reset a forgotten PIN. Now using the code, will return the phone to its original state, losing most forensic data. If the code is entered incorrectly 10 times in a row, the device becomes permanently blocked and unrecoverable.

ICCID: Integrated Circuit Card Identification Each SIM is identified by its integrated circuit card identifier (ICCID). These numbers are engraved on the SIM during manufacturing. This number has sub sections that are very important for forensics. Starting with the Issuer identification number (IIN), which is a seven-digit number that identifies the country code and issuer. There is also a variable length individual account identification number to identify the specific phone, and a check digit.

Cell on Wheels (COW) is a term for telecom infrastructure placed on a trailer to facilitate temporary expansion of cellular service. This can be used in emergency situations, particularly when natural disasters can take out existing cell towers and simultaneously increase demand for cellular service. As an example, in

2004 in the aftermath of Hurricane Charlie, several COWs were deployed in Southwest Florida.

3GPP

3GPP, or 3rd Generation Partnership Project is actually a group of standards organizations who collaborate to develop standards for mobile communications. Anyone working with mobile device forensics should be familiar with 3GPP. This group created the standards for GSM, UMTS, LTE, and 5G, as well as other standards. The organizational partners are shown in Table 1.7.

In addition to the organizational partners, there are a number of market partners. These groups have input and provide market advice, but they do not have the ability to define or publish standards.

3GPP Phase 1 was released in 1992. Since then, there have been multiple releases updating the standards to cover new technologies. Release 99 in 2001 specified the first UMTS networks. Release 8 in 2009 was the first release of LTE. The next release is Release 17 in 2022 that will cover various changes to 5G.

There are a number of 3GPP standards. The documents are quite voluminous. If you wish to focus on the 3GPP Security Standards, this list should be of some assistance:

- UMTS Security:
  - 33.102 Security Architecture.
  - 33.105. 3GPP Cryptographic Algorithm Requirements.
  - 35.201. f8 and f9 Specification.
  - 35.202. KASUMI Specification.
- IMS Security:
  - 23.228 IMS Architecture.

TABLE 1.7
3GPP Organizational Partners

Organization	Country/Region
Association of Radio Industries and Businesses (ARIB)	Japan
Alliance for Telecommunications Industry Solutions (ATIS)	U.S.A.
China Communications Standards Association (CCSA)	China
European Telecommunications Standards Institute (ETSI)	Europe
Telecommunications Standards Development Society (TSDSI)	India
Telecommunications Technology Association (TTA)	South Korea
Telecommunication Technology Committee (TTC)	Japan

- LTE Security:
  - 33.401 System Architecture Evolution (SAE); Security architecture
  - 33.402 System Architecture Evolution (SAE); Security aspects of non-3GPP
- Lawful Interception:
  - 33.106 Lawful interception requirements
  - 33.107 Lawful interception architecture and functions
  - 33.108 Handover interface for Lawful Interception
- Key Derivation Function:
  - 33.220 GAA: Generic Bootstrapping Architecture (GBA)
- Backhaul Security:
  - 33.310 Network Domain Security (NDS); Authentication Framework (AF)
- Relay Node Security
  - 33.816 Feasibility study on LTE relay node security (also 33.401)
- Home (e) Node B Security:
  - 33.320 Home (evolved) Node B Security
- All documents available for free at: ftp://ftp.3gpp.org/specs

A5/3 Block cipher used in UMTS, GSM, and others. It was designed explicitly for the 3GPP Standard by Security Algorithms Group of Experts (SAGE), a part of the European standards body ETSI.

OTHER WIRELESS TRANSMISSIONS

The obvious focus when discussing transmissions on a mobile device is cellular transmission. And that has been the primary focus in this chapter. However, mobile devices are also capable of connecting to wireless internet and of communicating over Bluetooth. Therefore, there should be some coverage of these technologies.

IEEE 802.11 Standard

Radio-wave-based networks adhere to the 802.11 standard. The 802.11 standard consists of several subclassifications that will be described in this section. This is generally what is referred to when one is discussing Wi-Fi computer networking.

802.11a

802.11a is an older Wi-Fi standard. You are unlikely to encounter it today. The 802.11a standard operated at the 5-GHz frequency with a maximum data rate of 54 Mbps. An 802.11a device could also use lower data rates of 48 Mbps, 36 Mbps, 24 Mbps, 18 Mbps, 12 Mbps, 9 Mbps, and 6 Mbps. At the 5-GHz frequency, 802.11a networking devices were not susceptible to interference from devices that cause interference at the 2.4-GHz frequency range. Devices compatible with the 802.11a standard were incompatible with 802.11b and 802.11g devices. Also, 802.11a devices use a higher frequency than 802.11b or 802.11g devices. The higher frequency cannot penetrate materials such as building walls like the lower frequency devices can. This results in 802.11a devices having a shorter range when compared with 802.11b, 802.11g, and 802.11n devices.

802.11b

Although the 802.11a and 802.11b standards were developed at the same time, 802.11b was the first to be adopted by industry. The maximum data rate for 802.11b was 11 Mbps. When the highest rate cannot be achieved because of distance or radio interference, a lower rate is automatically selected. The lower rates are 5.5 Mbps, 2 Mbps, and 1 Mbps. An 802.11b device can operate over any of 11 channels within the assigned bandwidth. When communicating between wireless devices, all devices should use the same channel. When using devices from the same manufacturer, the same channel is automatically selected by default. Two wireless networks, one constructed of 802.11b devices and the other constructed of 802.11a devices, can coexist without interfering with each other because they use different assigned frequencies. This allows for two different wireless networks to operate within the same area without interfering with each other.

802.11g

The IEEE 802.11g standard is also an older standard that is rarely used today. It followed the 802.11a and 802.11b standards. The 802.11g standard operates in the 802.11b frequency range of 2.4 GHz. This made it backward compatible with 802.11b devices. When communicating with 802.11b devices, the maximum data rate was reduced to 11 Mbps. The maximum throughput for the 802.11g standard was 54 Mbps, but the maximum distance is typically much shorter than an 802.11b device. Since 802.11g was assigned to the same frequency range as 802.11b, it is susceptible to the same sources of radio interference.

802.11n

The 802.11n standard operates at either 2.4 GHz or 5.0 GHz. This dual band modality continues with later standards. 802.11n implemented MIMO technology, as have all the subsequent standards. Multiple-input-multiple-output (MIMO) is a wireless networking technology that uses two or more streams of data transmission to increase data throughput and the range of the wireless network. Transmitting two or more streams of data in the same frequency channel is referred to as spatial multiplexing.

802.11n incorporates the multiple-input-multiple-output (MIMO) technology using 5-GHz and 2.4-GHz frequencies with an expected data rate of approximately 300 Mbps to 600 Mbps. The exact speed depends on the number of simultaneous data streams transmitted. Some 802.11n devices are advertised with data rates much higher than specified in the standard.

802.11n 2009

As the name suggests, IEEE 802.11n 2009, is an amendment to 802.11n. This standard describes technology that achieves bandwidth of up to 600 Mbit/second with the use of four spatial streams at a channel width of 40 MHz. It uses MIMO, which uses multiple antennas to coherently resolve more information than possible using a single antenna.

802.11ax

There have been several iterations of 802.11ax, each with its own advantages. These iterations include the following:

- IEEE 802.11-2012: This standard basically combined the improvements from 2007 to 2012 into a single standard.
- IEEE 802.11ac. This standard was approved in January 2014 and has a throughput of up to 1 Gbps with at least 500 Mbps and uses up to 8 MIMO.
- IEEE 802.11ad. This standard was developed by the Wireless Gigabyte Alliance and supports data transmission rates up to 7 Gbit/second.
- IEEE 802.11af. Approved in February 2014, 802.11af allows WLAN operation in TV white-space spectrum in the VHF and UHF bands between 54 and 790 MHz. It is also referred to as White-Fi and Super Wi-Fi.
- 802.11-2016. This revision incorporated 802.11 ae, aa, ad, ac, and, af into a single standard.
- IEEE 802.11aj. This is a rebranding of 802.11ad for use in the 45-GHz unlicensed spectrum available in some regions of the world, specifically China.
- 802.11aq. This is an amendment to the 802.11 standard to enable pre-association discovery of services. It does not affect bandwidth or transmission speed.
- 802.11ax This standard is meant to replace 802.11ac. The goal is to increase the throughput of 802.11ac. This standard was approved in February 2021 and is often marketed as Wi-Fi 6.
- 802.11ay This standard is still being developed as of this writing. It is intended to be an extension of 802.11ad to extend throughput and range.

802.11 Channels

Today you are probably using some variation of 802.11ax. In addition to the standard of wireless access point you use, the channels used are also important. The 802.11 standard defines 14 channels. The channels that can be used are determined by the host nation. In the United States, a WAP can only use channels 1 through 11. Channels tend to overlap, so nearby WAPs should not use close channels. For example, two nearby WAPS using channels 6 and 7 are likely to have interference issues.

In some cases WAPs can use channel bonding. Channel bonding is a method whereby two or more links are combined. This is done either for redundancy, fault tolerance, or for increased throughput. Channel bonding can be used in wired or wireless networks.

Wi-Fi Security

There have been four primary protocols for security Wi-Fi transmissions. Each is described here in chronological order.

Wired Equivalent Privacy (WEP) was the first method for security wireless networks. It uses a robust stream cipher, RC4. However, the implementation was flawed leading to serious security issues. WEP should simply not be used today.

Wi-Fi Protected Access (WPA) is a protocol that combines authentication with encryption. It uses Temporal Key Integrity Protocol (TKIP). TKIP is a 128-bit per-packet key, meaning that it dynamically generates a new key for each packet.

Wi-Fi Protected Access 2 (WPA2) was developed by the Wi-Fi organization as an enhanced version of WPA. WPA2 completely implemented the IEEE 802.11i security standard. It provides the following: The Advanced Encryption Standard (AES) using the Counter Mode-Cipher Block Chaining Message Authentication Code Protocol (CCMP). This provides data confidentiality, data origin authentication, and integrity for wireless frames.

Wi-Fi Protected Access 3 (WPA2) was released in January 2018 as a replacement to WPA2. WPA3 can use AES 256 bit in Galois Counter Mode with SHA-384 bit as a HMAC. This provides substantially more security than WPA 1 or WPA 2. WPA3 also requires attackers to interact with your Wi-Fi for every password guess they make, making it much harder and time-consuming to crack passwords. One of the important new security features is that with WPA3, even open networks will encrypt your individual traffic.

Bluetooth

Bluetooth is a short-range, wireless system that is designed for limited distances. Many texts and courses teach that Bluetooth has a maximum range of 10 meters. However, that is only partially true. In fact, it is only true for Bluetooth 3.0. The following table summarizes the ranges and bandwidth for the various versions of Bluetooth.

Ranges and Bandwidth for Bluetooth

Version	Bandwidth/Range
3.0	25 Mbit/second 10 meters (33 ft)
4.0	25 Mbit/second 60 meters (200 ft)
5.0	50 Mbit/second 240 meters (800 ft)

Bluetooth uses 79 separate channels that use the frequency-hopping spread spectrum transmission technique, starting at 2.4 GHz. The Bluetooth standard was developed separately from the IEEE network standards.

Bluetooth 5.2 was published in December of 2019. It adds some features, but not additional bandwidth or transmission ranges. Among the features are that audio will be transmitted using Bluetooth Low Energy. Bluetooth Low Energy, or BLE has been available since 2006. The purpose of BLE is to provide Bluetooth range and bandwidth while consuming less energy, as the name suggests.

CHAPTER SUMMARY

This chapter has provided you with the fundamentals of mobile networks. We have covered the basics of electromagnetic radiation as well as how to handle multiple access to a given signal. The basics of cellular networks have been covered including network types and the components of mobile networks. This material is fundamental to understanding how mobile networks function. That knowledge, in turn, forms a basis for studying mobile device forensics.

CHAPTER 1 ASSESSMENT

1. Which of the following has the longest wavelength?
   a. Radio waves
   b. Gamma Rays
   c. Ultraviolet
   d. X-Rays
2. _____ works by using an allocated band of frequencies and changing between the frequencies, using one frequency at a time.
   a. FHSS
   b. DSSS
   c. CSS
   d. THSS
3. The term _____ means there are multiple separate radio channels side by side within a designated radio band
   a. Multiplexing
   b. Duplexing
   c. Modulation
   d. Orthogonal
4. _____ uses the entire bandwidth to broadcast a signal, and relies on a sinusoidal signal of frequency increase or decrease.
   a. FHSS
   b. DSSS
   c. CSS
   d. THSS
5. _________requires the use of guard bands.
   a. FDMA
   b. TDMA
   c. CDMA
   d. QDMA

6. When using ______ orthogonal codes are used to separate different transmissions. Each symbol to be transmitted is encoded using a specific code.
   a. FDMA
   b. TDMA
   c. CDMA
   d. QDMA
7. Pico cells cover _____
   a. Home or office
   b. Small urban zones
   c. Mid-size area
   d. A building or tunnel
8. _____ This is the part of the cell network responsible for communications between the mobile phone and the network switching system.
   a. BTS
   b. MSC
   c. VLR
   d. BSS

2 Mobile Hardware and Operating Systems

INTRODUCTION

In chapter 1 you were given an overview of mobile networks, including how signals are encoded. This chapter will introduce you to the hardware found on mobile devices as well as the operating systems used for mobile devices. Frequently forensic examiners attempt to engage in forensics without this basic knowledge. Instead, they depend on some tool (i.e., Cellebrite, Mobile Edit, Oxygen, etc.) to give them the information they need. This is a serious mistake. Failure to understand the mobile device hardware and operating systems can lead to an inability to understand the limits of a given tool. No tool is perfect. In some instances, you will want to go beyond a tool and manually seek evidence. That is not possible without sufficient understanding of the hardware and software. One of the themes of this book is that the forensic examiner should have a thorough understanding of the hardware, operating system, network, and in fact all aspects of the device being examined. In the case of this book, that is mobile devices.

HARDWARE

As you are undoubtably aware, cellular phones have limited space for hardware, thus getting hardware small, is an important factor. Ultimately, however, the modern smartphone is simply a small computer. As such it must have similar components to what one finds in a computer. As one clear example, the smartphone must have a central processing unit (CPU). However, unlike your typical PC, the CPU in a smartphone will usually be integrated with a system on a chip process.

System on a chip (SoC) processors, as the name suggests, integrate many different functions into a single integrated circuit. Normally this will include CPU, memory, secondary storage. It may also include a graphics processing unit (GPU), Wi-Fi, and other technologies. This is in contrast to the personal computer that uses a motherboard with various circuits (including CPU, memory, and secondary storage) attached to it.

There are different types of SoC built for different purposes. Some are built around a microcontroller. Other SoC's are built around an application-specific integrated circuit or a field programmable gate array. The SoC's used in mobile phones are usually built around a microprocessor.

DOI: 10.1201/9781003118718-2

All SoC must have at least one processor core, but often they have more than one processor core. Multi-Core SoC are common today. Particularly with mobile devices, the SoC may also include a digital signal processor. Digital signal processing is needed to convert from the digital format of the mobile device to the analog format of the electromagnetic waves used for communication. The mathematics of a DSP can be rather complex but will be covered later in this chapter. The purpose of a DSP is to measure real-world analog signals then compress or filtering them. This process is used to translate between analog and digital.

Antenna

Clearly antennas are a vital part of cellular communications. Antennas are used to radiate and receive EM waves (energy), Antennas consist of one or several radiating elements through which an electric current circulates. Antennas typically are characterized by directivity, radiation pattern, gain, and efficiency. In some sources you will see the term aerial used for antenna. The electrical engineering symbol for antenna/aerial is shown in Figure 2.1.

There are two primary classes of antenna. The omnidirectional antenna, as the name suggests radiates EM in all directions. Directional antennas radiate in a particular direction. There are subsets of each of these, which we will explore later in this chapter. However, all antennas can be classified into one of these two families.

The simplest and most widely used type of antenna is the dipole antenna. This is sometimes called a doublet. These antennas consist of an electric dipole that has a radiating structure so that the current has only one node at each end. Commonly dipole antenna consists of two conductors oriented end-to-end with a feedline between them. Dipole antennas are sometimes used as components in a more complex antenna arrangement.

There are variations of the dipole antenna. One such variation is the short dipole. This antenna uses two conductors whose total length is less than half the wavelength of the transmissions. In general there is a relationship between the length of

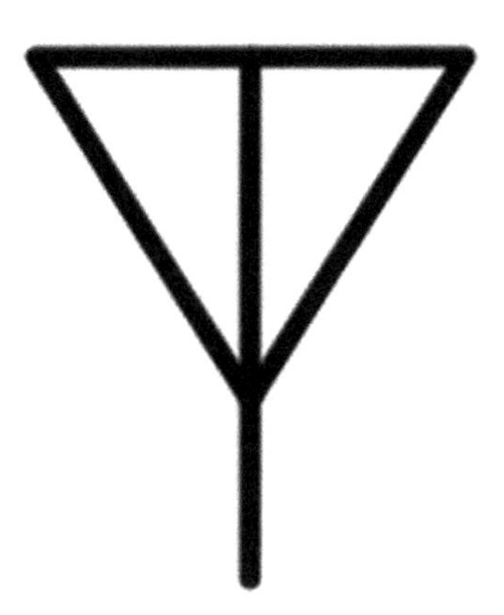

FIGURE 2.1 The electrical engineering symbol for antenna.

the dipole relevant to the wavelength of the transmission and the directive gain measured in dBi. The dBi unit is one of several antenna measurements. Common units are listed here:

- dBi: dB (isotropic): the forward gain of an antenna compared with the hypothetical isotropic antenna.
- dBd: dB (dipole): the forward gain of an antenna compared with a half-wave dipole antenna.
- dBiC: dB (isotropic circular): the forward gain of an antenna compared to a circularly polarized isotropic antenna.
- dBq: the forward gain of an antenna compared to a quarter wavelength whip.
- dB-Hz: bandwidth relative to one hertz.
- dBov or dBO: dB(overload): the amplitude of a signal compared with the maximum which a device can handle before clipping occurs.
- dBk: power relative to 1 kilowatt.

A concept found frequently in antenna engineering is the isotropic antenna. This is a theoretical point source of electromagnetic waves that radiates the same intensity in all directions. These are used as a theoretical comparator for actual antennas.

Principal characteristics used to characterize an antenna are:

- radiation pattern
- directivity
- gain
- efficiency

Gain is an important parameter in any antenna. This is defined as the ratio of the intensity radiated in a given direction at some distance, divided by intensity radiated at the same distance by a hypothetical isotropic antenna. The formula is shown in Equation 2.1:

$$G_{dbi} = 10\log\frac{I}{I_i} \tag{2.1}$$

The I_i is the intensity of the isometric antenna. Equation two provides decibels which are also called decibels -isotropic. Another united to measure gain is the decibels-dipole. This is used for half-wave dipole antenna. That formula is described in Equation 2.2.

$$G_{dBd} = 10\log\frac{I}{I_d} \tag{2.2}$$

The radiation pattern of an antenna is a plot that shows the radio waves relative strengths at different angles. One way this is viewed is via a Hertzian dipole. This is another theoretical antenna used as a comparison for actual antennas. Figure 2.2 shows the radiation pattern of a simple Hertzian dipole.

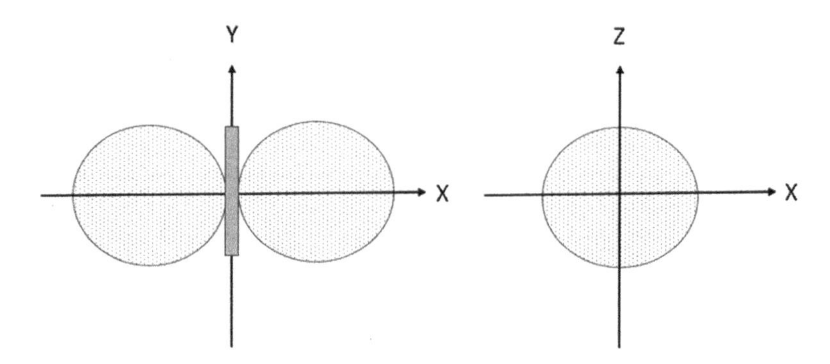

FIGURE 2.2 The Hertzian dipole radiation pattern.

One of the fundamental properties of real antennas is that the receiving pattern is identical to the far-field pattern used for transmitting. This fact comes from the reciprocity theorem of electromagnetics. There is actually a family of reciprocity patterns used in electromagnetism. The details of these are not germane to our exploration of mobile device hardware.

The space around an antenna is usually divided into three concentric regions. These are the reactive near-field, the radiating near-field, and the far field. You will sometimes encounter alternative names for these. The reactive near-field is sometimes called the inductive near-field; the radiation near-field is sometimes called the Fresnel region; and the far field is sometimes called the Fraunhofer region.

As an example of how these different fields affect the transmission consider the far field region. In the far-field radiated power decreases as the square of the distance. In the near field regions, the distance is negligible, but there is absorption of the radiation, and can even cause magnetic induction.

This discussion of these regions/fields might seem to be a diversion into more electrical engineering than the forensic examiner requires. However, this has a direct impact on antenna size. If the antenna is shorter than ½ the wavelength of the radiation emitted, the far and near regions are measured by a simple ration of the distance r from the source to the wavelength of the radiation.

Another important issue in antenna design is efficiency. No technology has 100% efficiency. The energy that does not actually go into the transmission usually turns into heat. For mobile devices, this can be a concern. There will always be some power loss due to impedance and the simple fact that no technology is perfect. But the level of efficiency in an antenna for a mobile phone affects not only battery life but also impacts overheating.

Free-Space Path Loss

With any electromagnetic signal, some signal will be lost even without interference. This is usually called free-space path loss. Free-space path loss is proportional to the square of the distance between the transmitter and receiver, and also proportional to the square of the frequency of the radio signal. A simplified version is shown in Equation 2.3.

$$\text{free space path loss} = \left(\frac{4\pi d}{\lambda}\right)^2 \tag{2.3}$$

d is the distance of the receiver from the transmitter (meters)
λ is the signal wavelength (meters)
f is the signal frequency (Hertz)
c is the speed of light in a vacuum (meters/seconds)

A more detailed understanding of free space path loss comes from the Friis transmission formula. This takes into account directivity of sending and receiving antenna. That formula is shown in Equation 2.4.

$$\frac{p_r}{p_t} = DtDr\left(\frac{\lambda}{4\pi d}\right)^2 \tag{2.4}$$

In Equation 2.4:

- P_r and P_t are the power received and the power transmitted.
- Dr and Dt are the directivity of the receiving and sending antenna.
- *d* is the distance between the antennae.
- λ is the wavelength of the transmission.

Equation 2.3 derives from Equation 2.4.

The Friis transmission formula bears a bit more discussion. It is quite important in telecommunications engineering. The formula was presented in 1946 by radio engineer Harald T. Friis. The formula originally published is shown in Equation 2.5.

$$\frac{P_r}{Pt} = \left(\frac{A_r A_t}{d^2\lambda^2}\right) \tag{2.5}$$

In Equation 2.5:

- P is the power fed into the transmitting or receiving antenna.
- A is the aperture of the receiving or transmitting antenna.
- *d* is the distance between the two antennas.
- λ is the wavelength of the transmission.

This eventually led to Equation 2.4, which in turn led to Equation 2.3. Thus, the current simplified formula for free space path loss traces back to the Friis formula from 1946.

Math behind DSP

The mobile phone also needs to have a digital signal processor. The essence of the processor is to convert the digital data of the mobile device into the analog electromagnetic signal, and vice versa. The circuitry in the DSP accomplishes this using mathematical formulas. There are two formulas often used.

The discrete cosine transform (DCT) is one such formula. This formula is used in many different applications. For example, it is used in digital video and audio. The formula was first proposed in 1972 by Nasi Ahmed. This function expresses a finite sequence of data points in terms of a sum of cosine functions oscillating at different frequencies. The DCT is shown in Equation 2.6.

$$X_k = \sum_{n=0}^{N-1} x_n \cos\left[\frac{\pi}{N}\left(n + \frac{1}{2}\right)k\right] \quad k = 0, \ldots, N - 1. \tag{2.6}$$

This function expresses a finite sequence of data points in terms of a sum of cosine functions oscillating at different frequencies. The N real numbers $x_0, \ldots, x_{N-1}$ are transformed into the N real numbers $X_0, \ldots, X_{N-1}$ using the transform. The DCT is a widely used transformation technique in signal processing and data compression.

The other formula is the Fourier transform. DSP's usually make use of the Fast Fourier Transform. However, to understand that, you must first understand the Fourier Transform. A Fourier transform basically decomposes a function of time (a signal) into the frequencies that make it up. The term Fourier transform refers to both the frequency domain representation and the mathematical operation that associates the frequency domain representation to a function of time. A Fourier transform essentially takes a function (one that can be expressed as a waveform) and decomposes it into its constituent frequencies.

What does the Fourier transform do? Given some wave function, it essentially extracts the frequencies present in that function. That may be an oversimplification, but it is a good place to start. Put a bit more precisely; a Fourier transform takes a function of time and transforms it into a function of frequency. Now let us take a look at the actual math. There are various forms of the Fourier transform, but here is one common one shown in Equation 2.7.

$$\hat{f}(\varepsilon) = \int_{-\infty}^{\infty} f(x)e^{-2\pi i x \xi}dx \tag{2.7}$$

For readers without a rigorous mathematical background, this might seem daunting. However, we will walk through each of the symbols, and even without a background in calculus, you should be able to grasp the essence of this formula.

First, the odd-looking f is a symbol for the Fourier transform itself. The symbol over the f is called a circumflex. The $\mathcal{E}$ represents some real numbers. This symbol is called the epsilon. Moving to the right side of the equation we encounter

$$\int_{-\infty}^{\infty}$$

For those readers without a background in calculus, this represents integration. An integral might be difficult to master, but the essence is quite easy. An integral basically takes small pieces of something and sums them up for a total. The classic example is the area under a curve. Imagine you have a curve, as shown in Figure 2.3, and want to know its area.

One way to compute that is to use smaller and smaller rectangular slices of it, summing up the areas of those slices, as shown in Figure 2.4.

As you get progressively narrower rectangles and sum their areas, your total area of the curve becomes more accurate. Now computing an integral does not actually involve adding up progressively narrower rectangles. In fact, it is because of

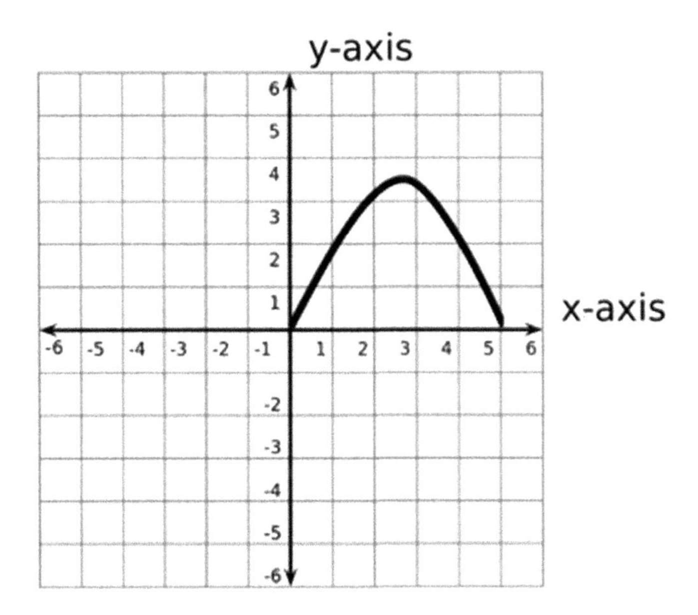

FIGURE 2.3 An example graph.

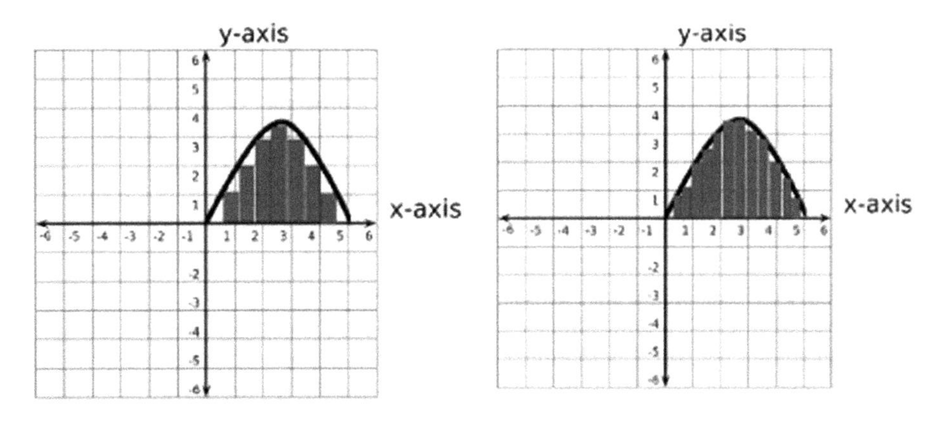

FIGURE 2.4 Breaking the graph down.

integrals that one does not have to do that. But that is the concept. The integral symbol is followed by the function one wishes to compute the integral of. This takes us back to what is remaining in the Fourier transform, shown in Equation 2.8.

$$f(x)e^{-2\pi i x \xi} \tag{2.8}$$

The function, *f(d)* is *e* raised to the power of $-2\pi i \times \mathcal{E}$. The *e* is Euler's number. This is usually approximated by 2.71828. The details of why this number is so important in wide ranging areas of mathematics is beyond the scope of this book. The 2π radians is 360 degrees. The final symbol dx is the symbol of differentiation. All integration is the integral of some differential. Differentiation deals with change. The process of differentiation is about finding the rate of change at a particular point. It is not inaccurate to think of integration and differentiation as opposite operations.

This brings us to Fast Fourier Transforms. There are actually quite a few variations of the FFT. We don't need to delve into those for you to understand the functionality of DSP's. It is merely useful for you to have a generalized understanding of Fourier Transforms.

This should provide you with a reasonable general understanding of the Fourier transform and the Discrete Cosine Transform. It is not critical that you have a deep understanding of these two formulas. They are presented just to show you the mathematics used in digital signal processors.

SIM Cards

As you are undoubtedly already aware, SIM cards are rather central to mobile devices.

This is a circuit that stores the International Mobile Subscriber Identity (IMSI). Think of it as how you identify the phone. Many modern phones have removable SIM, which means you could change out the SIM and essentially have a different phone with a different number. A SIM card contains its unique serial number (ICCID), the IMSI, security authentication and ciphering information. This SIM will also usually have network information, services the user has access to and two passwords. Those passwords are the personal identification number (PIN) and the personal unblocking code (PUK).

The first SIM cards were specified by the European Telecommunications Standards Institute and built by Giesecke & Devrient, a German company that specializes in banknote securities and smart cards. The first 300 SIM cards were sold to a Finnish wireless network operator called Radiolinja. Their network also hosted the world's first GSM phone call on March 27, 1991. SIM cards were specified by the European Telecommunications Standards Institute in specification TS 11.11. The standard is now governed by the 3GPP.

Before SIM cards, some phones used a NAM. Number Assignment Module stores telephone number, International mobile subscriber identity and an Electronic Serial Number, and other data. It essentially serves the same purpose as a SIM card but is a chip permanently part of the mobile device.

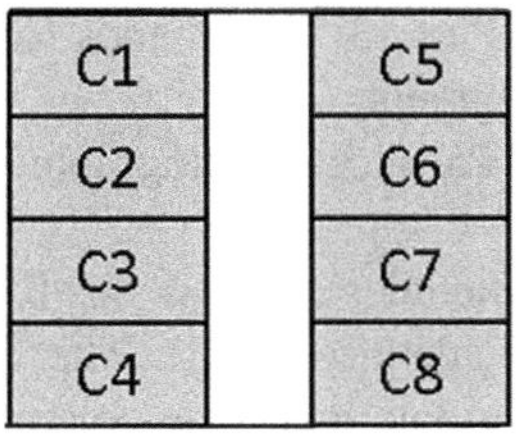

C1		C5
C2		C6
C3		C7
C4		C8

FIGURE 2.5 SIM card.

A very basic schematic of a SIM card is shown in Figure 2.5.

In Figure 2.5, the following are the various components:

- C1: Supply voltage (1.8, 3, 5 volts DC).
- C2: Reset signal
- C3: Clock signal (1 to 5 MHz, external)
- C4: Reserved
- C5: Ground
- C6: Programming voltage
- C7: Input/output Baud rate is (clock frequency)/372.
- C8: Reserved for SIM communication

The three voltages shown (1.8, 3, and 5 v) related to class A, B, and C SIM cards based on the ISO/IEC 7816 standard. There are several parts to this standard, all specifying, with substantial detail, the aspects of the SIM card. Those subparts are listed here:

- ISO/IEC 7816-1:2011 Part 1: Cards with contacts – Physical characteristics
- ISO/IEC 7816-2:2007 Part 2: Cards with contacts – Dimensions and location of the contacts
- ISO/IEC 7816-3:2006 Part 3: Cards with contacts – Electrical interface and transmission protocols
- ISO/IEC 7816-4:2013 Part 4: Organization, security and commands for interchange
- ISO/IEC 7816-5:2004 Part 5: Registration of application providers
- ISO/IEC 7816-6:2016 Part 6: Interindustry data elements for interchange
- ISO/IEC 7816-7:1999 Part 7: Interindustry commands for Structured Card Query Language (SCQL)
- ISO/IEC 7816-8:2016 Part 8: Commands and mechanisms for security operations
- ISO/IEC 7816-9:2017 Part 9: Commands for card management
- ISO/IEC 7816-10:1999 Part 10: Electronic signals and answer to reset for synchronous cards.
- ISO/IEC 7816-11:2017 Part 11: Personal verification through biometric methods
- ISO/IEC 7816-12:2005 Part 12: Cards with contacts – USB electrical interface and operating procedures

- ISO/IEC 7816-13:2007 Part 13: Commands for application management in a multi-application environment
- ISO/IEC 7816-15:2016 Part 15: Cryptographic information application

Data stored on a SIM card includes at least the ICCID code, the IMSI, the authentication key, and a local identity (LAI). The card may also include carrier specific data, and possibly user data including contacts. The ICCID is probably the most significant from a forensic perspective.

ICCID is an acronym for Integrated Circuit Card Identification Integrated Circuit Card Identifier. Each SIM is identified by its integrated circuit card identifier (ICCID). These numbers are engraved on the SIM during manufacturing. This number has subsections that are very important for forensics. This number starts with the issuer identification number (IIN), which is a seven-digit number that identifies the country code and issuer, followed by a variable-length individual account identification number to identify the specific phone, and a check digit. The ICCID has some interesting information. From ForensicWiki[1] the example 89 91 10 1200 00 320451 0

The first two digits (89 in the example) refers to the Major Industry Identifier.
The next two digits (91 in the example) refers to the country code (91-India).
The next two digits (10 in the example) refers to the issuer identifier number.
The next four digits (1200 in the example) refers to the month and year of manufacturing.
The next two digits (00 in the example) refers to the switch configuration code.
The next six digits (320451 in the example) refers to the SIM number.
The last digit which is separated from the rest is called the checksum digit.

Country codes can be 2 or 3 digits. The codes are defined by the ITU-T in standards E.123 and E.164. These are the same codes used for international dialing. So, for example, Chile would be 56, China 86, Japan 81, etc.

The International mobile subscriber identity (IMSI) is the next most important information on a SIM card.

- The first 3 digits represent the mobile country code (MCC).
- The next 2 to 3 digits represent the mobile network code (MNC).
- The next digits represent the mobile subscriber identification number (MSIN). The number of digits in the MSIN varies. However, the standard dictates that the total length of the IMSI should be less than 15 digits.

The SIM card also includes the location area identity (LAI). This is a unique identifier that consists of the three-digit mobile country code (MCC) and the 2 to 3-digit mobile network code (MNC) and a location area code (LAC) that identifies a specific area. Sim cards come in several formats, Table 2.1 summarizes those.

TABLE 2.1
SIM Card Formats

SIM Card Format	Introduced	Length	Width	Thickness
Full-size (1FF)	1991	85.6 mm /3.37 in	53.98 mm /2.125 in	0.76 mm /0.030 in
Mini-SIM (2FF)	1996	25 mm /0.98 in	15 mm /0.59 in	0.76 mm /0.030 in
Micro-SIM (3FF)	2003	15 mm /0.59 in	12 mm /0.47 in	0.76 mm /0.030 in
Nano-SIM (4FF)	2012	12.3 mm /0.48 in	8.8 mm /0.35 in	0.67 mm /0.026 in
Embedded-SIM (eSIM)	2010	N/A	N/A	N/A

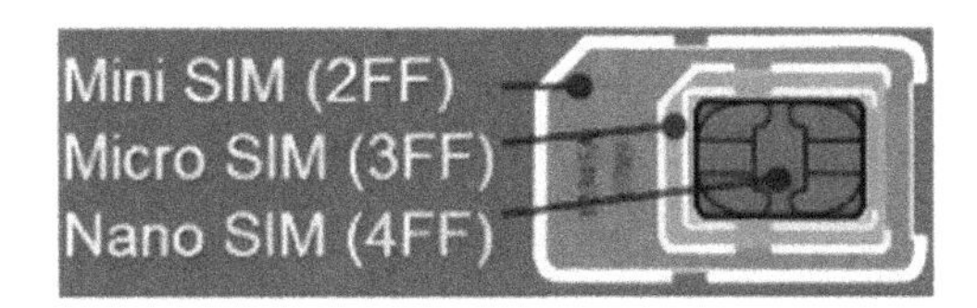

FIGURE 2.6 NIST SIM cards.

SIM standards can be found ETSI standard GSM 11.11.[2] The United States National Institute of Standards (NIST) NIST SP 800-101 Guidelines on Mobile Device Forensics describes SIM cards in the following manner:

"UICCs are available in three different size formats. They are: Mini SIM (2FF), Micro SIM (3FF), and Nano SIM (4FF). The Mini SIM with a width of 25 mm, a height of 15 mm, and a thickness of .76 mm, is roughly the footprint of a postage stamp and is currently the most common format used worldwide. Micro (12 mm × 15 mm × .76 mm) and Nano (8.8 mm × 12.3 mm × .67 mm) SIMs are found in newer mobile devices (e.g., iPhone 5 uses the 4FF)." Figure 2.6 is from the NIST standard.

The NIST standard also describes the file system used on a SIM card. That file system has three elements: the root of the file system (MF), subordinate directory files (DF), and files containing elementary data (EF). Figure 2.7 is from the NIST Standard.

CPU

Any computing device must have a central processing unit. Mobile phones are no different. Much of our discussion of CPU's would also be applicable to a CPU found in a laptop, personal computer, or similar device. The CPU is often described as the "brains" of any computer device, and that is an accurate description. It is in the CPU that the processing of data actually takes place. The essential functions a CPU must accomplish are shown in this list:

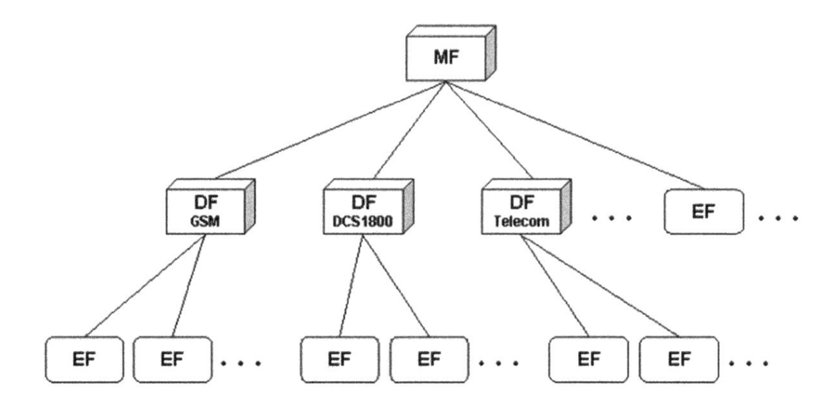

FIGURE 2.7 NIST SIM file system.

- CPU fetches instructions from memory.
- CPU interprets instructions to determine action that is required.
- CPU fetches data that may be required for execution (could come from memory or I/O).
- CPU processes data with arithmetic, logic, or some movement of data.
- CPU writes data (results) to memory or I/O.

A general overview of CPU components is shown in Figure 2.8

The control unit, as the name suggests, directs the operation of the processor. It is directing the other subunits. The arithmetic logic unit performs math using binary logic operations. The memory management unit translates the logical addresses used by the CPU to actual physical memory addresses. Which brings us to the address generation unit. This calculates addressed used by the CPU to access the main memory.

CPU's also have registers. Registers are essentially the CPU's small amount of memory for it to use directly, apart from the main memory of the device. These are normally small, such as 32-bit or 64-bit registers. There are registers specific for data (called data registers). Then there are address registers to hold addresses from main memory. There are registers specifically for decimal numbers, called floating point registers, and there are general purpose registers.

Mobile devices often used ARM Architecture. This acronym has changed over time. It once meant Acord RISK machine or Advanced Risk Machine. Whatever

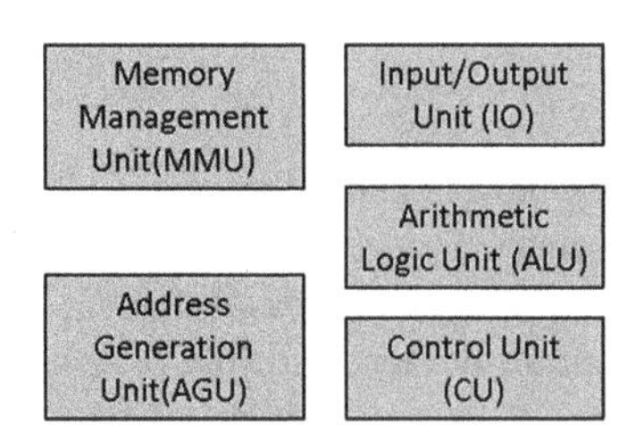

FIGURE 2.8 CPU architecture.

one takes the acronym to mean it is really about a specific architecture for CPU's. They are RISC processors. The RISC acronym is less ambiguous, it stands for Reduced Instruction Set Computer. The term reduced is intended to describe the fact that the amount of work any single instruction accomplishes is reduced. Not necessarily there are fewer instructions. There are other CPU designs such as CISC (Complex Instruction Set) and MISC (Minimal Instruction Set). RISC processors average 1 instruction per clock cycle.

One example of an ARM architecture that is used in some mobile phones is the ARM v8-A. The general characteristics of that CPU are:

- It uses 64-bit addresses.
- It has 31 general purpose 64-bit registers.
- The instructions, however, are still primarily 32-bit instructions.
- Supports double-precision floating point.
- IT also has hardware acceleration for cryptography.

OTHER DEVICES

Aside from the hardware directly related to mobile equipment, there are other devices that are related to the topic of mobile telephony. Some such devices have very significant ramifications for mobile device security and forensics.

JAMMERS

As the name suggests, these are devices intended to disrupt cellular communication. In many countries, including the United States, these devices are illegal. The Communications Act of 1934, as amended, and the Commission's rules do not permit the use of transmitters designed to prevent or jam the operation of wireless devices in hospitals, theaters and other locations. Section 302(a) of the Communications Act, 47 USC 302(a), prohibits the manufacture, importation, sale, offer for sale, or use of devices that fail to comply with the regulations promulgated pursuant to this section.

Based on the above, the operation of transmitters designed to jam wireless communications is a violation of 47 USC 301, 302(a), and 333. The manufacture, importation, sale or offer for sale, including advertising, of such transmitters is a violation of 47 USC 302(a). Parties in violations of these provisions may be subject to the penalties contained within 47 USC 501-510. Fines for a first offense can range as high as $11,000 for each violation or imprisonment for up to one year. The equipment can also be seized and forfeited to the U.S. Government. These regulations apply to all transmitters that are designed to cause interference to, or prevent the operation of, other radio communication systems.

These devices are also illegal in the EU, India, Canada, United Kingdom, and many other countries. Most nations have exceptions for specified law enforcement usage. In some countries it is legal to own, but illegal to use a jammer. In other countries it is illegal to even own such a device. It is not recommended that you experiment with these devices or attempt to construct your own homemade jammer. These devices function using a transmitter that is tuned to the same frequency and modulation method (see chapter 1) as the target device. This allows the device to "jam" the target device.

IMSI Catchers

The most well-known IMSI catcher is the StingRay. It essentially is a fake cell phone tower. Phones connect to it and then the call can be monitored. At one point the existence of these devices were classified. They are still limited in use to specific law enforcement purposes with a valid warrant. There are handheld versions with brand names such as KingFish. There have even been reports of body worn IMSI catchers.

IMSI catchers operating in passive and active modes. The passive mode is simply analyzing traffic. The active mode is emulating a cell tower. Some of the active mode uses are:

- Extracting stored data such as International Mobile Subscriber Identity ("IMSI") numbers from phones that connect to the IMSI catcher (thus the name, IMSI catcher).
- Altering transmission signal power
- Forcing a device to downgrade to an older, less secure communication protocol.
- Intercepting the communications, or at least the metadata.
- Being able to locate the connected mobile device.

As you might imagine, there are a wide range of counter measures for IMSI catchers. The efficacy of these is not clear. One such app that can be found in the Apple App store or the Google store is the Cell Spy Catcher, you can see this in Figure 2.9.

SOFTWARE DEFINED RADIO

Radio is traditionally implemented by hardware components. Software Defined radio implements the components primarily in software running on a standard computer with perhaps some additional hardware. The ideal receiver scheme would be to attach an analog-to-digital converter to an antenna. A digital signal processor would read the converter, and then its software would transform the stream of data from the converter to any other form the application requires.

Software-Defined Radio (SDR) refers to the technology wherein software modules running on a generic hardware platform consisting of DSPs and general-purpose microprocessors are used to implement radio functions such as generation of transmitted signal (modulation) at transmitter and tuning/detection of received radio signal (demodulation) at receiver. A basic SDR system may consist of a

FIGURE 2.9 SpyCatcher.

personal computer equipped with a sound card, or other analog-to-digital converter, preceded by some form of RF front end.

Software Communications Architecture is a standard for SDR. It includes the following:

- This is an open standard for radio communication. It is widely used by SDRs. It was first published by the Joint Tactical Networking Center.
- The Standard Waveform APIs define the key software interfaces that allow the waveform application and radio platform to interface.
- The Core Framework (CF) defines the essential "core" set of open software interfaces and profiles that provide for the deployment, management, interconnection, and intercommunication of software application components in an embedded, distributed-computing communication system.

This is the conceptual goal that Software Defined Radio is moving towards.

A cognitive radio (CR) is a radio that can be programmed and configured dynamically to use the best wireless channels in its vicinity to avoid user interference and congestion. This concept was first proposed by Joseph Mitola III in 1998. The first cognitive radio wireless regional area network standard was

TABLE 2.2
Radio Types

Tier	Name	Description
Tier 0	Hardware Radio (HR)	Implemented using hardware components. Cannot be modified
Tier 1	Software Controlled Radio (SCR)	Only control functions are implemented in software: interconnects, power levels, etc.
Tier 2	Software Defined Radio (SDR)	Software control of a variety of modulation techniques, wide-band or narrow-band operation, security functions, etc.
Tier 3	Ideal Software Radio (ISR)	Programmability extends to the entire system with analog conversion only at the antenna.
Tier 4	Ultimate Software Radio (USR)	Defined for comparison purposes only

IEEE 802.22 published in 2011. Different types of radio are shown in Table 2.2.

There are some basic components that an SDR must have. These are shown in the following list:

- Analog Radio Frequency (RF) receiver/transmitter in the 200 MHz to multi-gigahertz range.
- High-speed A/D and D/A converters to digitize a wide portion of the spectrum at 25 to 210 M samples/second.
- High-speed front-end signal processing including Digital Down Conversion (DDC) consisting of one or more chains of mix + filter + decimate or up conversion.
- Protocol-specific processing such as Wideband Code Division Multiple Access (W-CDMA) or OFDM, including spreading/de-spreading, frequency-hop-and chip-rate recovery, code/decode functions, including modulation/demodulation, carrier and symbol rate recovery, and channel interleaving/de-interleaving.
- Data communications interface with carrier networks and backbone for data I/O and command-and-control processing, usually handled by general purpose ARM or PowerPC processors and Real-Time Operating System (RTOS).

CHAPTER SUMMARY

Chapter 2 expanded on what you learned in chapter 1. The essential hardware used in mobile devices was explored, in some detail. Antenna, DSP, CPU, and SIM cards were described. Some of the mathematics related to digital signaling was also described. This includes Discrete Cosine Transform, Fast Fourier Transforms, and the

Free-space Path Loss formula. Other related devices were also described. Jammers and IMSI catchers were briefly described. You also were given an introduction to software-defined radio.

CHAPTER 2 ASSESSMENT

1. What best describes an isometric antenna?
   a. An omnidirectional antenna
   b. A directional antenna
   c. An SDR antenna
   d. A theoretical ideal antenna
2. What is a concern with near field regions of antenna that is not a major concern with the far field region?
   a. Absorption of the radiation
   b. Distance attenuation
   c. Free-space path loss
   d. Power attenuation
3. What does the following equation describe?

$$\frac{P_r}{Pt} = \left(\frac{A_r A_t}{d^2 \lambda^2}\right)$$

   a. Free-space path loss
   b. Friis equation
   c. Discrete Cosine Transform
   d. Fast Fourier Transform
4. Which of the following are voltages used with SIM cards?
   a. 1
   b. 1.8
   c. 3
   d. 3.8
   e. 5
   f. 5.8
   g. 7
5. The second two digits in an ICCID represent what?
   a. Country code
   b. Network identifier
   c. SIM number
   d. Issuer identifier

6. Where should you look to find the mobile network code (MNC)?
   a. First two digits of the ICCID
   b. Digits 4–6 of the IMSI
   c. First two digits of the IMSI
   d. Digits 4–6 of the ICCID

NOTES

1 https://forensicswiki.xyz/wiki/index.php?title=SIM_Cards
2 https://www.etsi.org/deliver/etsi_gts/11/1111/05.03.00_60/gsmts_1111v050300p.pdf

3 iOS Operating System

Apple Devices are quite common, thus the iOS operating system is an important aspect of mobile forensics. This includes fundamental facts of the operating system, where to look for data, and related facts. An understanding of the history of iOS is also important. Perhaps of most interest to forensic practictioners are issues related to iOS security.

When performing mobile forensics, the most likely operating systems you will encounter are iOS and Android. There are certain tools that will extract data for you. However, no tool is perfect. It is also the case that you may need to circumvent the security of the device in order to perform the digital forensic examination. For this reason, it is critical to understand the details of the operating system involved. In this chapter we will explore the iOS operation system in depth.

OPERATING SYSTEM FUNDAMENTALS

Before delving into the details of iOS, it is critical that you have a working understanding of operating systems. This material will also be repeated, with a few small variations, in chapter 4: *Android Operating Systems*. There are issues that apply to all operating systems that are important when trying to understand a particular operating system, such as iOS.

The purpose of an operating system is to provide a means of controlling, managing, and communicating with the system's hardware, and other software on the system. Due to the iOS operating system, a programmer does not write an app for specific hardware. He or she simply writes the app to work with iOS. The operating system (iOS) then takes care of managing resources and communicating with the hardware.

Because of all the functions that operating systems perform, they also provide a framework for other applications to work in. A given program need not know exactly how to access a file on a drive or allocate memory for a variable. That program simply requests those functions from the operating system. It should be clear that the operating system is the most important program on any given computing device, including an iPhone.

There have been numerous operating systems throughout the years. Operating systems can be single tasking (performing only one task at a time) or multi-tasking. Operating systems used for end user; general purpose devices are always multi-tasking. There are also operating systems that support multiple simultaneous users or a single current user. Embedded operating systems are designed to work in embedded devices. These usually have a narrower set of capabilities.

All operating systems start with the kernel. You may think of the kernel as the core of the operating system. It is one of the first pieces of software loaded when the

DOI: 10.1201/9781003118718-3

system powers on, immediately after the bootloader. The kernel will interact at a low level with memory, storage, and other components. In operating system parlance, low level refers to how close to actual machine instructions something is. So low-level means close to the actual machine instructions being executed on a given chip. High-level refers to code that is removed from the machinery, such as a user program.

Kernels will handle some basic low-level functionality such as memory management and multitasking. The kernel is loaded into a separate part of memory referred to as kernel space. This memory space is not accessible by the user or by applications. Usually, an operating system will have an application programming interface (API) that facilitates communication between the user space and kernel space.

Discussion of user space and kernel space brings us to the topic of memory management. One of the other core elements of any operating system is memory management. Every program has to have a space in memory set aside in which it will run. Larger, more complex programs may only have part of the program in memory with other modules loaded and unloaded as needed. However, all programs require memory to be allocated. The allocation of memory is a function of the operating system. The operating system must also deal with the situation of a program, or one of its components, being unloaded from memory. Another aspect of memory management is memory protection. It is critical that one program not be able to access anything in another program's memory space. Keeping programs within their own memory space or process and preventing them from interfering with other programs is a function of the operating system.

There are two approaches to kernel design. Those are micro-kernel and monolithic kernel. With a monolithic kernel design many operating system services run in the same memory area along with the kernel itself. Microkernels have a small kernel with bare minimum functionality, then other services such as network connectivity, are provided in separate memory in user memory space.

These are the two fundamental approaches to kernel design. There are other less commonly used kernel design approaches. The nanokernel is one, with an even smaller kernel than microkernel. There are also modular kernels (sometimes called Hybrid kernels) that are widely used in commercial operating systems. Apple macOS uses a hybrid kernel called XNU, which we will discuss in detail later in this chapter. Windows NT based systems use a hybrid kernel. To illustrate the difference between hybrid kernels and micro or monolithic kernels, let us focus on XNU. XNU is an acronym for X is Not Unix. XNU itself is a heavily modified version of the open-source Mach kernel. It runs the core of the operating system it hosts as separate processes (the modular design approach). In general, a hybrid or modular kernel will have multiple components in the kernel mode space. This can include the Hardware Abstraction Layer (HAL), Kernel mode drivers, and a microkernel. There is then an executive layer between the kernel mode and the user mode. This executive layer contains things such as process managers, input/output management, power management, graphics management, etc. The hybrid kernel approach is shown in Figure 3.1.

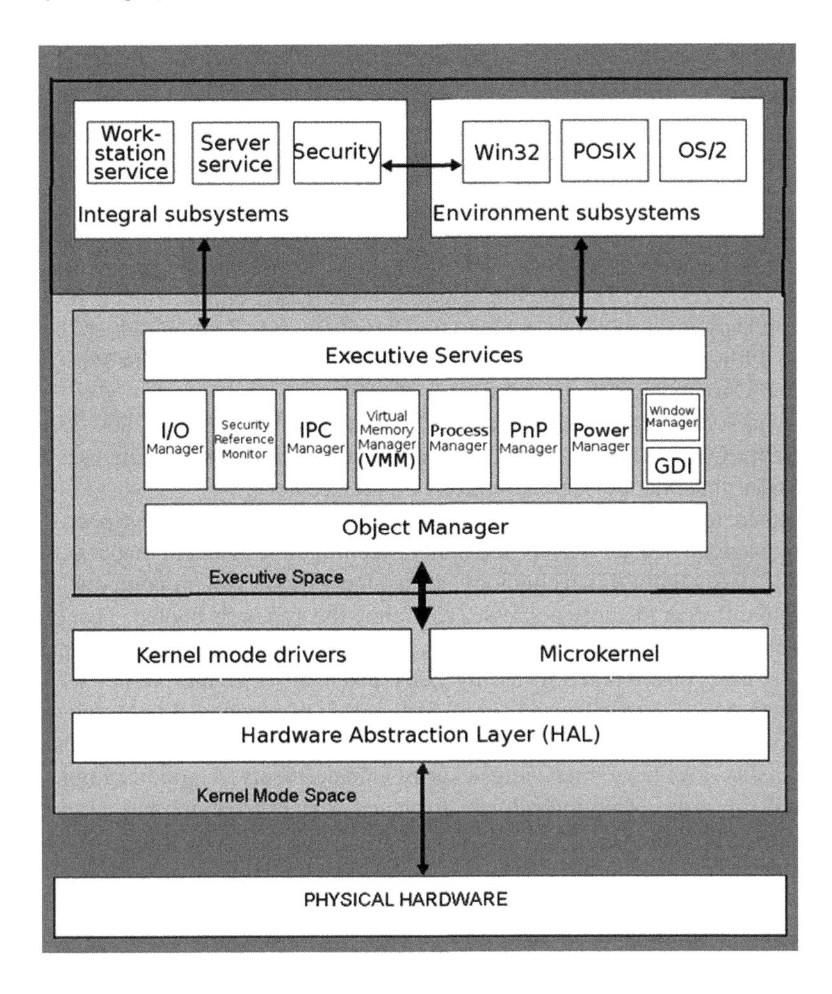

FIGURE 3.1 The hybrid kernel approach.

Understanding operating system kernels has direct implications for mobile device forensics. Very few forensics techniques or tools can access data in kernel space. The major exception is JTAG, which we will be discussing in chapter 7: *JTAG and Chip-off.* However, this is not a significant concern. The suspect, and any apps he or she is using, also are very unlikely to be able to access kernel space. Therefore, it is very unlikely you will find much evidence there, even if you can access it.

In addition to the operating system itself, there is the file system. A file system organizes the data on the file. Remember that storage devices are storing data either as sectors on a traditional hard drive, or as cells on a solid-state drive. And the data at the drive level is either magnetic or electrical. The file system is how such raw data is organized into files and directories. NTFS, as an example is a file system used by Microsoft. We will examine the APFS file system used by iPhone and iPad later in this chapter.

IOS BASICS

The iOS operating system is used by iPhone, iPod, and iPad. It was originally released in 2007 for the iPod Touch and the iPhone. The user interface is based all on touching the icons directly. It supports what Apple calls gestures: swipe, drag, pinch, tap, and so on. The iOS operating system is derived from OS X. It should be noted that the computer/laptop version of Macintosh is now simply called MacOS.

The iOS kernel is the XNU kernel of Darwin. The original iPhone OS (1.0) up to iPhone OS 3.1.3 used Darwin 9.0.0d1. iOS 4 was based on Darwin 10. iOS 5 was based on Darwin 11. iOS 6 was based on Darwin 13. iOS 7 and iOS 8 are based on Darwin 14. iOS 9 is based on Darwin 15. iOS 10 is based on Darwin 16. iOS 11 is based on Darwin 17. iOS 12 is based on Darwin 18.

Darwin is an open-source Unix code first released by Apple in 2000. Darwin is the core for OS X, iOS, watchOS, tvOS, etc. The iOS operating system uses 256-bit encryption; thus, the device encryption is quite secure.

Another feature of the XNU kernel is its use of Address Space Layout Randomization (ASLR). This is a feature of memory management that is not unique to Apple. Essentially this technology causes the various system components to be loaded to different memory addresses each time the system is booted. That prevents an attacker or malware from exploiting known memory addresses. The term ASLR was first used in reference to a July 2001 patch to the Linux kernel which implemented ASLR. Apple began using ASLR with iOS version 4.3 in March of 2011.

As of this writing iOS 14 is the current version, it was released in June 2020. This version does have interesting security enhancements. A good example is that with iOS 14, a recording indicator is displayed whenever an app has access to the microphone or camera.

There are four layers to iOS. The first is the Core OS layer. This is the heart of the operating system. Next is the Core Services layer. The Core Services layer is how applications interact with the iOS. Next is the Media layer, which is responsible for music, video, and so on. Finally, there is the Cocoa Touch layer, which responds to the aforementioned gestures. The four layers are often diagramed as shown in Figure 3.2.

The Core OS layer has lower-level processes that are needed by the system. For example, the core Bluetooth framework is found in the Core OS layer. Also in the Core OS layer is the security services framework and the local authentication framework.

The core services layer has all the various services you might expect. For example, the address book framework, core location framework, cloud kit framework, core motion framework, and healthkit framework are found in this layer. This is the layer that iOS apps frequently interact with.

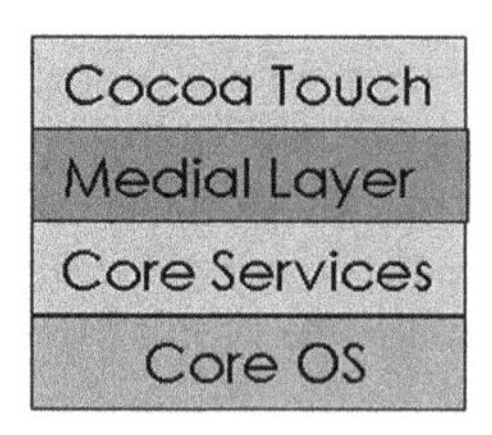

FIGURE 3.2 The iOS layers.

As the name suggest, the media layer is responsible for all the various multimedia functionality. This includes the UIkit Graphics used by app developers. It also includes the core graphics, images, and animation frameworks. There is also the Metal api. This is an application programming interface first provided with iOS 8. It provides hardware accelerated 3D graphics functionality. The media layer also has the AV Kit and AV Foundation for audiovisual.

The Cocoa touch layer is where user haptics are processed into system commands. There is also the EventKit framework and the MapKit frameworks at this layer. App developers work extensively with the Cocoa touch layer.

The file system for iOS has a similar structure to Unix. The root directory is called root and is denoted by simply/. Usually, a user cannot access this. Under the root there are system data and user data. Fortunately, the criminals also cannot access system data, thus only accessing user data is not prohibitive for iOS forensics. A summary of the various iOS versions is provided in Table 3.1.

Obviously, Table 3.1 does not list every feature of every single version. The objective is to familiarize the reader with the general features in each version of iOS. And of course, one would expect new versions to come with additional new features. Two trends are clear. The first is improved interactivity support with features such as AirDrop, QR Code reading, etc. The second trend is a constant push for a more secure phone. This can be a problem for forensic examiners who might need to circumvent security (with a warrant of course) when a suspect refused to provide access to the device.

Since iOS 10.3, iOS has used the Apple File System (AFS). This is the same file system used on Macintosh 10.13 and later, as well as tvOS 10.2 and watchOS 3.2. It is a replacement for the older HFS+. AFS supports full disk and file encryption. It also supports read-only snapshots of the file system at a point in time. AFS supports 64-bit inode numbers allowing a much larger number of files to be stored. This file system also uses checksums to ensure metadata data integrity.

The iOS contains several elements in the data partition:

- Calendar entries
- Contacts entries
- Note entries
- iPod_control directory (this directory is hidden)
- iTunes configuration
- iTunes music

This exact list can vary from model to model.

The iOS runs on iPhones, iPods, and iPads. This means that once you are comfortable with the operating system on one Apple device, you should be comfortable with any Apple device. This applies not just to the features that users interact with, but also to the operating system fundamentals. Thus, if you have experience with forensics on an iPhone, you will have no problem conducting a forensic analysis of an iPad.

If you are unable to access the iPhone, it is possible to get much of the data from the suspects iCloud account. As of 2018, 850 million customers back up to the cloud.

TABLE 3.1
iOS History

Version	Basic Facts
iOS 4	Released to the public in June 2010. This version has not been supported since December 2013.
iOS 5	Released to the public in October 2011. iCloud storage and wireless updates were introduced with this version of the iOS. This was also the first version of iOS with the camera app.
iOS 6	Released to the public September 2012. This version introduced Siri, the personal assistant. It also came with an integrated Facebook app, as well as the new "do not disturb setting". Version 6.0 also saw the new Apple maps replace the Google maps. This was also the first version to have the Passbook app to hold boarding passes, tickets, coupons, etc.
iOS 7	Released to the public September 2013. This was a substantial change in the user interface. It also added the AirDrop wireless sharing technology to the iOS. The notification center was overhauled and CarPlay (formerly iOS in the car) was also introduced in iOS 7.1.
iOS 8	Released to the public in September 2014. One of the major enhancements was Continuity, which allows cross platform sharing between iPhone, iPad, and Macintosh devices. This version also included HealthKit that allows programmers to write health-related apps. The Weather app now uses the Weather Channel instead of Yahoo.
iOS 9	Released to the public in September 2015. This version included updates to the battery and the iOS takes up less space in storage than previous versions. iOS 9 also introduced six-digit pass codes and support for two factor authentication. This version also used the Metal API for improved speed and performance.
iOS 10	Released to the public September 2016. Moved from slide to unlock to press the home screen. If liquid is detected in the Lightning port, the user is warned. With iOS 10.2 the TV app is included by default.
iOS 11	Released to the public in September 2017. A more human voice for Siri and support for other languages. QR scanning is added for the camera app.
iOS 12	Released to the public in September 2018. Performance was optimized. The ARKit 2 allows users to share their view with other iOS 12 devices.
iOS 13	Released to the public in September 2019. General upgrades to performance, interface, and Siri. Battery performance was improved substantially. The new Core Haptics was included to allow developers to provide better haptics in apps. The iPad version was named iPadOS.
iOS 14	Released to the public in September 2020. Car keys allows the iPhone to be a virtual car key using NFC technology. There were improvements to CarPlay and the home screen. Now uses Wi-Fi MAC address randomization, providing each Wi-Fi hotspot with a different Mac address. There is also a new sandbox feature called "BlastDoor" that is used with untrusted data in iMessage.

Information stored in the iCloud can be retrieved by anyone without having access to a physical device, provided that the original Apple ID and password are known. Often time the username and password can be extracted from the user's computer.

The tool 3uTools is downloadable for free from http://www.3u.com/. You can see the main screen in Figure 3.3.

FIGURE 3.3 3uTools for iOS.

You can see it immediately displays a great deal of important information such as IMEI, iOS version, SPU, etc. You can also get a full listing of apps on the phone as well as music, photos, and other data. You have to jailbreak the iPhone to extract call and message history.

JAILBREAKING

The topic of jailbreaking is unavoidable when exploring the iOS. It is a process of escalating privileges to remove restrictions on the iOS. This is very similar to the rooting process in Android which we will explore in chapter 4 *Android operating system.* At one time jailbreaking was required to support using the iPhone as a Wi-Fi hotspot, a process known as tethering. That is no longer the case as the iOS supports tethering. However, people still sometimes jailbreak the phone in order to install programs not approved by Apple. Another reason for jailbreaking is to unlock the phone from a particular carrier.

There have been various utilities developed for aiding in jailbreaking an iOS device. RedsnOw, uncOver, Absinthe, and Pangu are a few. Jailbreaking should not be undertaken lightly. At a minimum it will circumvent security mechanism yielding a far less secure device. It is also possible to damage the iOS or filesystem so substantially that the phone is no longer usable.

FILE SYSTEM

For several years, Apple has used its proprietary file system named Apple File System (APFS) for its products. This has been the case since macOS High Sierra (10.13), tvOS 10.2, and iOS 10.3. Apple had previously used the HPFS+ file system which has in turn supplanted the HFS file system (Hierarchical File System). APFS is designed for solid state storage and flash drives.

APFS uses 64-bit inode numbers. An inode is a method for file systems to track file portions. The inode approach is also used in the ext file system, very popular with Linux distributions. An inode is a data structure that describes a file or directory. Each inode has information such as last time accessed, last time modified, owner, and permission data. This is how some file systems such as APFS keep track of files and directories. Because APFS is using 64-bit inodes it can support a very large number of files, with each address being 64 bits long.

APFS was first announced in 2016. It is designed so that it can be used even on devices with small storage space. APFS uses the GPT partition scheme. GPT, or GUID Partition Scheme uses globally unique identifiers (GUID's) for identifying blocks of data. Blocks of data are addressed using LBA or logical block addressing. This is a common approach now used with many systems other than APFS. GPT is a method of organizing files that is used on a number of devices, including standard computer hard drives and solid-state drives. The GPT approach is part of the Unified Extensible Firmware Interface (UEFO) standard used with PCs.

The GPT scheme uses one or more APFS containers and each container may have one or more volumes. APFS volumes support snapshots that are point in time read-only instances of the file system. These are useful in recovering from some system failures. APFS also supports file compression. There are three different compression algorithms supported: LZFSE (Lempel-Ziv Finite State Entropy), LSVN, and Deflate. All three are called Lempel-Ziv compression algorithms. Lempel Ziv algorithms work by replacing repeated occurrences of data elements with a single copy and a pair of numbers that indicate where the next repletion of that data element will occur.

There are several features of APFS that have direct impact on digital forensics. The first is the support of full disk encryption. This can be a substantial challenge for the forensic examiner. There are three modes of disk encryption supported:

- No encryption
- Single key encryption
- Multi key encryption where each file is encrypted with a separate key

It is important to keep in mind that the presence of encryption does not mean it is impossible to get at the data. The likelihood of being able to crack the encryption algorithm is virtually zero, and you will see why in chapter 10 Anti Forensics. However, there are also ways to get around encryption that will also be discussed in chapter 10. There are also occasionally flaws in the implementation of encryption. For example, in January 2021 an APFS driver used in iOS 14.4 was found to have a bug that would allow one to read arbitrary files, regardless of their permissions.

APFS also supports snapshots that are read-only instances of the file system at some particular point in time. These can provide forensic evidence, even in cases where the suspect has attempted to delete evidence.

HARDWARE DETAILS

The touchscreen has changed substantially over the years. For the first several years there was a 3.5-inch liquid crystal displace (LCD). Later versions of the iPhone allowed for larger sizes. For example, the iPhone XS Max and 11 Pro Mad have a 6.5-inch screen. Resolution has also improved. Early iPhone offered 320×480 resolution. By iPhone 4S this had improved to 640×960. iPhone 4 also changed to a different type of LCD called an in-plane switching LCD. The iPhone X was the first to use the OLED display.

OLED is an acronym for Organic Light Emitting Diode. This can provide a fare superior quality of image than LCD. There are two types of OLED, the passive matrix (PMOLED) or active matrix (AMOLED). With PMOLED each row is controlled sequentially. With AMOLED each individual pixel is controlled, thus leading to a much better resolution.

For several years, iPhone has used a coprocessor named Secure Enclave. This processor handles cryptographic keys and biometric information. This coprocessor has its own secure boot process.

There are a number of variations on the OLED structure. The first variation is whether the device is bottom or top oriented. This refers to the direction that light exits the device. Bottom devices light passes through the bottom electrode and substrate that the panel was manufactured on. Top devices light exists through the lid that is added during the manufacture of the OLED device. Top OLED's are often used for active-matrix applications. There are also stacked OLED's. These are so named because the red, green, and blue sub pixels are stacked rather than side by side.

The iPhone 6s introduced the 3D touch displays. These displays not only recognize touch, but the pressure of the touch. This facilitates improved haptics. The device can tell the difference between a light touch and a firm touch. But with the advent of iPhone 11, the 3D touch was replaced with a newer technology called Haptic Touch. The Haptic Touch cannot detect differences in pressure. These devices look at touch duration rather than pressure.

Sensors

iPhones have a number of sensors. These can be accessed by Apple apps, the iOS itself, or by third party apps. There are proximity sensors to determine if the device is brought near the face during a call. Ambient light sensors, as the name suggests, adjust the display brightness based on ambient light. Magnetometers have been part of iPhone since iPhone 3GS. These measure strength and direction of the magnetic field and are useful in geolocation.

Accelerometers sense the orientation of the phone. This was originally done to alter the screen to match the current orientation. However, the sensor can be used for many other applications, such as detecting elevation and speed. Starting with iPhone 5S, there is an M7 motion coprocessor included to improved motion detection and sensitivity. Related to the magnetometer is the barometer. This was first introduced with iPhone 6 and determines air pressure, and thus elevation.

Gyroscopes are used to measure orientation and angular momentum. Taken together: accelerometer, gyroscope, magnetometer, and barometer can provide a very accurate measurement of location, speed, and momentum.

Various biometric sensors have been added over time. The fingerprint sensor was first added to iPhone 5S. The iPhone X added the TrueDepth camera system for accurate facial recognition. These technologies are integral to the advanced security features available on current models of the iPhone.

iPhone Processor

Apple has used ARM-based system on a chip (SoC0 designed by Apple inc. for several years. Beginning with the iPhone 8 the Apple A11 Bionic chip was used. The A11 is a six core CPU with two high-performance cores and four energy-efficient cores. Perhaps most interesting is that the A11 includes hardware dedicated to supporting artificial neural networks. Apple calls this a "neural engine". The newer iPhone 12 uses the Apple A15 Bionic system on a chip. The basic design is still four energy-efficient cores with two high-performance cores, like the A11. However, the A14 has a faster CPU and graphics processing unit. The iPhone 12 also has improved the neural engine so that it is now a 16-core neural engine with machine learning matrix accelerators. Table 3.2 provides an overview of the various processors, including security features.

iPHONE SECURITY

The area that may be of most interest to forensic examiners is the area of security. As was mentioned previously, one may have a warrant to examine a phone, but the suspect simply does not provide access. In chapter 10 we will explore countermeasures, however in this section you will see a general overview of iPhone security measures.

Apple devices, including the iPhone have a cryptographic processor. Keys for encryption are stored there. iPhone uses AES encryption with a 256-bit key. This is quite strong encryption. Strong enough that public court cases have indicated the United States FBI cannot breach it. After the 2015 San Bernardino terrorist attack, the FBI was unable to get into the phone of the suspects and took Apple to court to force them to unlock the phone. This demonstrates that generally speaking a forensic examiner is unlikely to be able to get around iPhone encryption. There is a tool called GrayKey which we will be discussing in chapter 10 that may have some success on this issue. It should be noted that with iPhone the encryption is not optional. When your phone is locked it is encrypted.

The iPhone has increased the length of passcode to 10, and now supports facial recognition as an option. The facial recognition is facilitated with the Secure Neural Engine. This is part of the Secure Enclave, which will be discussed in detail later in this chapter. Beginning with the A11 system on a chip, the Secure Neural Engine was integrated into the Secure Enclave. However, starting with

Table 3.2
Overview of iPhone Processors

SoC	Memory Protection Engine	Chip Details	Cryptography
A8	Encryption and authentication	64-bit 1.4 GHz.	AES Engine
A9	Encryption and authentication	64-bit ARM (Advanced RISC Machines) based 1.85 GHz.	DPA (Dynamic Power Analysis) protection, PKA (Public Key Accelerator)
A10	Encryption, authentication, and replay prevention	64-bit ARM based. Two cores each 2.34 GHz. Two additional low-performance cores. With a total of four cores.	DPA protection and lockable seed bits, OS-Bound Keys
A11	Encryption, authentication, and replay prevention	64-bit ARM based. Two cores each 2.39 GHz. Two additional low-performance cores. With a total of four cores.	DPA protection and lockable seed bits, OS-Bound Keys
A12	Encryption, authentication, and replay prevention	64-bit ARM based. Quad core with two high-performance cores at 2.49 GHz.	DPA protection and lockable seed bits, OS-Bound Keys
A12 (after Fall 2020)	Encryption, authentication, and replay prevention	64-bit ARM based. Eight core with two high-performance cores at 2.49 GHz.	DPA protection and lockable seed bits, OS-Bound Keys
A13	Encryption, authentication, and replay prevention	64-bit ARM based. Six core with two high-performance cores at 2.65 GHz. Also has a four core GPU (Graphics Processing Unit).	DPA protection and lockable seed bits, OS-Bound Keys, Secure Boot Monitor
A14	Encryption, authentication, and replay prevention	64-bit ARM based. Six core with two high-performance cores called Firestorm and four energy-efficient cores called Icestorm. Also has 16-core neural engine.	DPA protection and lockable seed bits, OS-Bound Keys, Secure Boot Monitor

A14 the Secure Neural Engine was moved to become part of the application processors neural engine.

Perhaps the most important is Apples Secure Enclave. This is an entire security subsystem that is integrated into Apples system on a chip. The Secure Enclave is isolated from the main processor, so that even if that should become compromised, the secure data is not in danger. During boot up there is a process to establish a hardware root of trust. The Secure Enclave also contains an AES encryption engine to facilitate cryptographic processes. This was first introduced with the iPhone 6s and has been a part of all iPhones since that time. It is also available on iPad, MacBook, Apple TV, and Apple Watch.

The Secure Enclave includes a root cryptographic key that is unique to each device and unrelated to any other identifiers on the device. At manufacturing a randomly generated UID (User ID) is fused into the system on a chip (SoC) processor. The Secure Enclave Processor (SEP) has its own operating system called sepOS. The Secure Enclave also has a device group ID (GID), which is common to all devices that use a given SoC.

The operations of the Secure Enclave processor are executed in a dedicated and protected region of the device's memory. Each time the system is booted the Secure Enclave generates a random ephemeral key for the Memory Protection Engine. Any data that the Secure Enclave writes to its dedicated memory region is encrypted using AES and uses a Cipher Based Message Authentication Code (CMAC) for integrity.

Beginning with the Apple A13 system on a chip, the Secure Enclave includes a boot monitor to ensure the integrity of the operating system being booted. Also, at startup the Secure Enclave uses its System Coprocessor Integrity Protection (SCIP) to prevent the Secure Enclave process from executing any code other than the legitimate Secure Enclave Boot Rom. What all of this means is that the Secure Enclave even protects the boot process, including the booting of its own chip.

Facial recognition has been a large part of the iOS security since the iPhone X. There is a separate facial recognition subsystem named Face ID. The hardware associated with Face ID includes a sensor as well as a projector that projects a grid of small infrared dots on the users face, and an illuminator that shines infrared light at the user. Then an infrared camera takes a photo of the user reading the resulting pattern. This allows the creation of a 3-D facial map. The Face ID system is advanced enough that adding or removing glasses, growing or shaving facial hair, and similar changes don't affect the systems ability to recognize the user.

CHAPTER SUMMARY

This chapter has covered the fundamentals of the iOS operating system, as well as iPhone hardware. This information forms the foundation of understanding iPhone forensics. While forensics tools will automatically produce much of this information for you, the evidence is more understandable if you have a strong working knowledge of the iOS operating system and of the phone hardware.

CHAPTER 3 ASSESSMENT

1. What file system does the iPhone use?
   a. HFS
   b. HFS+
   c. EXT
   d. APFS

2. What kernel is used in iOS?
   a. Micro kernel
   b. Monolithic
   c. XNU
   d. Darwin

3. _____ causes the various system components to be loaded to different memory addresses each time the system is booted.
   a. Micro kernel
   b. ASLR
   c. APFS
   d. Address space layer randomization

4. Where is the core motion framework found?
   a. Core OS
   b. Touch layer
   c. Kernel
   d. Core services

5. Which version of iOS introduced 6-digit passcodes?
   a. iOS 9
   b. iOS 10
   c. iOS 11
   d. iOS 12

6. What is the primary signal used in Apple Face ID?
   a. Visible light
   b. Ultraviolet light
   c. Infrared light
   d. Radio waves

CHAPTER 3 LABS

For these labs you will need an Apple phone. Even an older model, or one with defects such as a broken screen will be adequate. As long as it powers on, that is the issue. Such phones can be found used on E-Bay, Amazon.com, and phone repair stores for approximately $20 US Dollars. If you have an iPhone you can use that. These labs will not hurt your device.

Lab 3.1

Use the free version of http://www.3u.com to extract as much data as you can from your iPhone.

4 Android Operating System

Android has a large percentage of the mobile device market, making it important for understanding mobile forensics. This knowledge must include the fundamentails of Android, including the history of the operating system. But it also includes more detailed information such as the various file systems and the system architecture. Furthermore, there are various tools for working with Android including Android Debugging Bridge (ADB).

ANDROID BASICS

Android is a very common operating system. It is obviously found on Android phones, but it is also found in smart TVs, automobiles, and some IoT devices. It is clearly quite important to understand the Android operating system in some depth. The Android operating system is a Linux-based operating system, and it is completely open source. If you have a programming and operating systems background, you may find it useful to examine the Android source code from http://source.android.com/.

Android was first released in 2003 and is the creation of Rich Miner, Andy Rubin, and Nick Sears. Google acquired Android in 2005, but still keeps the code open source. The versions of Android have been named after sweets:

- Version 1.5 Cupcake
- Version 1.6 Donut
- Version 2.0-2.1 Éclair
- Version 2.2 Froyo
- Version 2.3 Gingerbread
- Version 3.1 - 3.2 Honeycomb
- Version 4.0 Ice Cream Sandwich
- Version 4.1 - 4.2 Jellybean
- Version 4.3 Kitkat
- Version 5.0 Lollipop released November 3, 2014
- Version 6.0 Marshmallow released October 2015
- Version 7.0 Nougat was released in August 2016
- Version 8.0 Oreo released August 2017
- Version 9.0 Pie August 6, 2018
- Android 10. This was called Q during beta testing, and it marks a departure from using the names of deserts. This was released in September 2019
- Android 11 Released February 2020

DOI: 10.1201/9781003118718-4

The differences from version to version usually involve adding new features, not a radical change to the operating system. This means that if you are comfortable with version 7.0 (Nougat), you will be able to do forensic examination on version 9.0 (Pie). While the Android source code is open source, each vendor may make modifications. This means even the partition layout can vary. However, there are common partitions that are present on most Android devices (phones or tablets).

The boot loader partition is necessary for hardware initialization and loading the Android kernel. This is unlikely to have forensically important data.

The boot partition has the information needed to bootup. Again, this is unlikely to have forensically important data.

The recovery partition is used to boot the phone into a recovery console. While the partition may not have forensically relevant data, sometimes you may need to boot into recovery mode.

The userdata partition is the one most relevant to forensic investigations. Here you will find the majority of user data including all the data for apps. We will discuss SQLite databases in chapter 8, this directory will contain many of those database.

The cache partition stores frequently accessed data and recovery logs. This can be very important for forensic investigations. There can be data here that the user is not even aware of. This means the user is less likely to have deleted it.

The system partition is not usually important for forensic examinations.

Remember that Android is Linux based, if you have an image of an Android phone you may be able to execute at least some Linux commands on it. For example, using cat proc/partitions will reveal to you the partitions that exist on the specific phone you are examining. In addition to these partitions there are specific directories that may yield forensic evidence.

- The acct directory is the mount point for the control group and provides user accounting.
- The cache directory stores frequently accessed data. This will almost always be interesting forensically.
- The data directory has data for each app. This is clearly critical for forensic examinations.
- The mnt directory is a mount point for all file systems and can indicate internal and external storage such as SD cards. If you have an Android image, the Linux ls command used on this directory will show you the various storage devices.

As you encounter dates on files in Android, there can be some confusion. Android can use several different date/time formats. The following lists the various formats available:

- UNIX, based on Jan 1, 1970
- GPS, based on Jan 6, 1980
- Some (such as Motorola) may use AOL, based on Jan 1, 1980

Fortunately, there are a number of online converters for date and time formats. You can simply search for a converter, but here are a few you may find useful:

https://www.unixtimestamp.com/index.php
https://timestamp.online/
https://www.onlineconversion.com/date_time.htm

SPECIALIZED KEY CODES

There are a number of keycodes that can be entered on an Android phone to get useful information from the phone. Some of these codes work with all Android models, others are specific to particular models. These are useful in diagnostics as well as forensics.

Diagnostic configuration *#9090#
Battery Status *#0228#
System dump mode *#9900#
Testing Menu *#*#4636#*#*
Display Info about device *#*#4636#*#*
Factory Restore *#*#7780#*#*
Camera Information *#*#34971539#*#*
Completely Wipe device, install stock firmware *2767*3855#
Quick GPS Test *#*#1472365#*#*
Wi-Fi Mac Address *#*#232338#*#*
RAM version *#*#3264#*#*
Bluetooth test *#*#232331#*#*
Displays IMEI number *#06#
Remove Google account setting *#*#7780#*#*
Toggle always-on display on or off *#99#
Log test settings *#800#
Engineering switch test mode *#801#
GPS TTFF (Time-To-First-Fix) test mode *#802#
Engineering Wi-Fi setting *#803#
Automatic disconnect test mode *#804#
Engineering bluetooth test mode *#805#
Engineering aging test mode *#806#
Engineering automatic test mode *#807#
Enter engineering mic echo test mode *#809#
Automatically searches for available TDSCDMA carriers *#814#
Automatically searches for available WCDMA carriers *#824#
Automatically searches for available LTE carriers *#834#
Automatically searches for available GSM carriers *#844#
Test photograph RGB (Red, Green, & Blue tint) *#900#
LCD display test *#*#0*#*#*
Packet loopback *#*#0283#*#*
Melody test *#*#0289#*#*

Proximity sensor test *#*#0588#*#*
Melody test *#*#0673#*#*
Test for vibration and backlight functionality *#*#0842#*#*
Advanced GPS testing *#*#1575#*#*
Touch screen test *#*#2664#*#*
Checks for root *#*#7668#*#*
Bluetooth test *#*#232331#*#*

These are just a sample of codes. One can search the internet for something like "Secret codes for XXXX phones" (replacing xxxx with the model you are interested in) to get information on codes for that specific phone model.

WORKING WITH ANDROID

Whether you are attempting diagnostics, performing a forensic exam, or testing an Android app in development, you will find the Android Debugging Bridge (ADB) to be useful. To extract data from an Android phone or tablet, it must be in developer mode. How you get there has changed with different versions, where to access developer mode is given here (note some models might have slightly different steps). Also note, that even if you are not using ADB, phone forensics tools require you to place the phone in developer mode to use the forensic software.

Settings> General> About and tap the Build number 7 times.

After tapping the Build Number 7 times, you will see a message "You are now a developer!"

Return to the main Settings menu and now you'll be able to see Developer Options.

Tap on Developer options and mark the box in front of USB Debugging to enable it.

To disable USB Debugging mode later, you can uncheck the box before the option.

To enable Developer Options, go to Settings> Developer options and tap on the ON/OFF slider on the top of the page.

Now that you can turn on and off the developer options, you can use the various phone forensics tools we will discuss in later chapters. However, you also can use the Android Debugging Bridge. As will be seen in chapter 6 this tool can be quite useful for performing Android forensics. However, it was actually written to provide debugging tools for Android developers, as the name suggests. It is also useful in performing basic Android phone diagnosis.

ADB

You can download this free tool from https://developer.android.com/studio/command-line/adb. The Android Debugging Bridge has three components:

- A client, which sends commands. The client runs on the development machine. The investigator can invoke a client from a command-line terminal by issuing an adb command.
- A daemon (adbd), which runs commands on a device. The daemon runs as a background process on each device.
- A server, which manages communication between the client and the daemon. The server runs as a background process on your development machine.

When you start an adb client, the client first checks whether there is an adb server process already running. If there isn't, it starts the server process. When the server starts, it binds to local TCP port 5037 and listens for commands sent from adb clients – all adb clients use port 5037 to communicate with the adb server. When a device is connected to a computer that has ADB, the first step is to list all connected devices. This is shown in Figure 4.1.

If the device does not show as attached, there are several possibilities. One common issue is that the developer mode is not turned on. Another possibility is that the ADB services has a problem. This requires you to first kill the service, then restart it as shown here:

adb kill-server

adb start-server

Once you have connected successfully to the Android device, there are several commands that are of interest when using ADB. One of the most interesting is the adb shell command. The adb shell command, enters a shell on the Android, from which the user can utilize standard Linux commands. This is shown in Figure 4.2

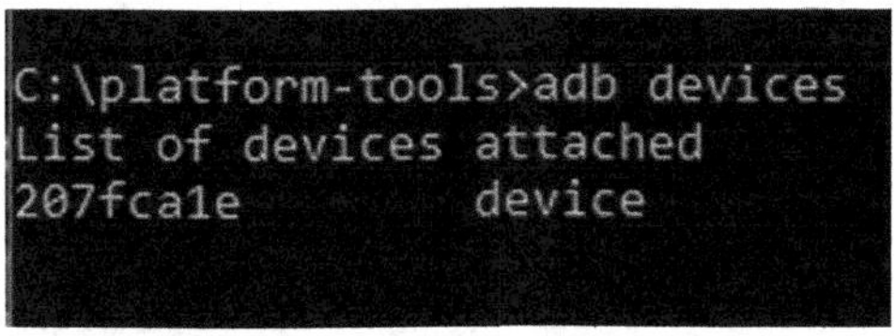

```
C:\platform-tools>adb devices
List of devices attached
207fca1e        device
```

FIGURE 4.1 ADB initial screen.

```
D:\platform-tools>adb shell
kanda:/ $ ls
acct        charger default.prop init                     init.usb.configfs.rc mnt  product sys
bin         config  dev          init.environ.rc          init.usb.rc          odm  sbin    system
bugreports  d       etc          init.rc                  init.zygote32.rc     oem  sdcard  ueventd.rc
cache       data    factory      init.recovery.m5621.rc   lost+found           proc storage vendor
kanda:/ $
```

FIGURE 4.2 ADB shell.

Common Linux commands done from within the adb shell include pstree, to list all the processes on the target device, in a tree format. Other standard Linux commands such as ps, ls, netstat, and lsof can also be useful. When you wish to exit the shell, just type exit as shown in Figure 4.3.

To see all the Linux commands available on a given device, from within the adb shell one can type ls/system/bin. This will show all the system binary files (i.e., executables). This is shown in Figure 4.4.

Aside from issuing Linux commands from within the adb shell, there are also adb specific commands that can be useful in a forensic examination. Common ADB commands are summarized in Table 4.1. Note, as you have seen in the screenshots,

```
C:\projects\teaching\Android\tools\platform-tools>adb shell
error: no devices/emulators found

C:\projects\teaching\Android\tools\platform-tools>adb shell
shell@y25c:/ $ exit

C:\projects\teaching\Android\tools\platform-tools>adb devices
List of devices attached
LGL16C9f11b9bf  device
```

FIGURE 4.3 ADB exit.

```
shell@y25c:/ $ ls /system/bin
ATFWD-daemon
PktRspTest
adb
adsprpcd
am
app_process
applypatch
atd
atrace
bdaddr_loader
blkid
bmgr
bootanimation
brctl
bridgemgrd
btnvtool
bu
bugreport
cat
charger_monitor
chcon
chmod
chown
clatd
clear
cmp
cnd
```

FIGURE 4.4 ls/system/bin.

TABLE 4.1
Common ADB Commands

Command	Purpose
adb pull	Pulls a single file or entire directory from the device to the connected computer.
adb restore <archive name>	Creates a backup of the device.
adb reboot	This causes the phone to reboot, there are several modes: adb reboot adb reboot recovery adb reboot bootloader
dumpsys	This is a very versatile command with several options: adb shell dumpsys package com.android.chrome will dump all the data for a given package. adb shell dumpsys activity provides information about Activity Manager, activities, providers, services, broadcasts, etc. adb shell dumpsys battery set level
pm list packages	This will list all packages on the device. There are a number of options for this command including: pm list packages -f See their associated file pm list packages -d Filter to only show disabled packages pm list packages -e Filter to only show enabled packages pm list packages -i See the installer for the packages pm list packages -u Also include uninstalled packages
dumpstate	adb shell dumpstate will dump a great deal of information. It is probably best to output this to a file, for example: adb shell dumpstate >state.txt
adb get-serialno	Print serial number to the screen
adb backup	Backup the phone: adb backup -apk -all -f backup adb restore backup.ab (restore a previous backup)
adb shell list packages	adb shell list packages -r (list package name + path to apks) adb shell list packages -3 (list third party package names) adb shell list packages -s (list only system packages) adb shell list packages -u (list package names + uninstalled) adb shell dumpsys package packages (list info on all apps)
adb shell install	adb shell install <apk> (install app) adb shell install <path> (install app from phone path) adb shell install -r <path> (install app from phone path) adb shell uninstall <name> (remove the app)
adb help	list all commands

you only have to type "adb shell" before a command if you are using a shell command and are not currently in the adb shell.

These commands combined with the specialized key codes discussed earlier can provide a great deal of information about the phone in question. This table is not an exhaustive list of adb commands, but it does provide the most common adb commands.

ROOTING ANDROID

Unfortunately, there is some data one cannot get to without rooting the Android phone. The term root is the Linux term for the administrator. In Linux, you simply type su (super user or switch user) and enter the root password. However, Android phones don't allow you to do that. Rooting a phone gives you complete root access to all aspects of the phone. However, that will also void any warranty.

In the past, rooting was not terribly difficult. There were even apps one could get that would root the phone for you. Most of these apps do not work on current models of Android. However, there are some methods that might work, depending on a number of variables. For example, the model you have, the version of Android, etc. will affect whether or not you will be successful. It is important to keep in mind that these are simply possible techniques. There is no guaranteed method for rooting an Android phone.

Before you can root a phone it must first be OEM (Original Equipment Manufacturer) unlocked. And it so happens, that before you can OEM unlock, you must first unlock it from the carrier. The only consistently effective method for carrier unlocking is to contact the carrier, assuming the contract is paid off, and request a carrier unlock. After that you have several options for OEM unlock, again depending on what manufacturer and model you are working with. If your phone has "OEM Unlock" enabled and visible under developer settings, then you have the option to use ADB (discussed in the previous section) and move to what is called fastboot mode. For most phones this is done by using ADB and typing in adb reboot bootloader. At that point you can try fastboot oem unlock. If that does not work, then your model requires you to get an unlock code and send it to the vendor to get OEM unlock. This is shown in Figures 4.5 through 4.7.

Once the phone is OEM unlocked, you can then root the phone. But be aware, that for many phones getting to the OEM unlock might be very difficult. by typing.

adb reboot bootloader

followed by

fastboot oem device-info

You will then see the devices OEM unlock status as shown in Figure 4.8.

Once it is OEM unlocked, you can use free tools such as Magisk to install a new image, that is rooted. Once you have installed Magisk you can update your

```
C:\Program Files (x86)\Minimal ADB and Fastboot>adb devices
List of devices attached
d79ede09        device

C:\Program Files (x86)\Minimal ADB and Fastboot>adb reboot bootloader

C:\Program Files (x86)\Minimal ADB and Fastboot>fastboot oem get_unlock_code
...
(bootloader) Unlock code:
(bootloader) ====================================
(bootloader) DE5C952C1564204CE112FC0DBD227E87
(bootloader) 05F6C0A4AF00EB0160EF86354EEE0F6A
(bootloader) ====================================
OKAY [  0.009s]
finished. total time: 0.010s

C:\Program Files (x86)\Minimal ADB and Fastboot>
```

FIGURE 4.5 OEM unlock the phone Step 1.

```
C:\Program Files (x86)\Minimal ADB and Fastboot>fastboot flash cust-unlock unlock_code.bin
target reported max download size of 536870912 bytes
sending 'cust-unlock' (0 KB)...
OKAY [  0.004s]
writing 'cust-unlock'...
(bootloader) Device is unlocked.
OKAY [  0.004s]
finished. total time: 0.009s
```

FIGURE 4.6 OEM unlock the phone Step 2.

```
C:\Program Files (x86)\Minimal ADB and Fastboot>fastboot oem unlock
...
OKAY [  0.036s]
finished. total time: 0.037s
```

FIGURE 4.7 OEM unlock the phone Step 3.

system. But you need to push the image onto the system for Magisk to be able to find it. That can be done with adb push. Also, for new images, they take up a lot of space, make sure you have enough space for the image. The process, once you have OEM unlocked is as follows. Once you have the phone OEM-Unlocked, rooting is not that difficult. You will need a TWRP image, those can be found at https://twrp.me/Devices/

- Place TWRP recovery image in “ADB and Fastboot” folder
- Open fastboot
- Check to make sure the device is connected with adb devices.
- Reboot in boot loader mode by adb reboot bootloader
- Use the phone screen to select Apply Update from ADB Sideload, or similar entry.
- type fastboot flash recovery twrp.img

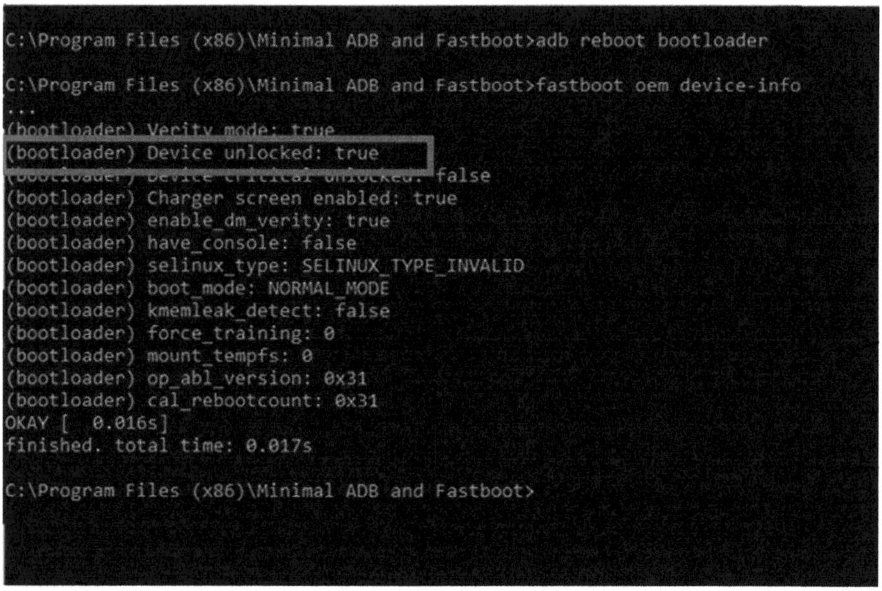

```
C:\Program Files (x86)\Minimal ADB and Fastboot>adb reboot bootloader

C:\Program Files (x86)\Minimal ADB and Fastboot>fastboot oem device-info
...
(bootloader) Verity mode: true
(bootloader) Device unlocked: true
(bootloader) Device critical unlocked: false
(bootloader) Charger screen enabled: true
(bootloader) enable_dm_verity: true
(bootloader) have_console: false
(bootloader) selinux_type: SELINUX_TYPE_INVALID
(bootloader) boot_mode: NORMAL_MODE
(bootloader) kmemleak_detect: false
(bootloader) force_training: 0
(bootloader) mount_tempfs: 0
(bootloader) op_abl_version: 0x31
(bootloader) cal_rebootcount: 0x31
OKAY [  0.016s]
finished. total time: 0.017s

C:\Program Files (x86)\Minimal ADB and Fastboot>
```

FIGURE 4.8 Verifying phone is unlocked.

Note that the twrp.img is whatever image you intend to put on the phone, and it must be in the fastboot directory. Also keep in mind, that with newer phone models, any one of these steps may fail. Rooting modern phones is a difficult task and frequently fails.

EDL MODE

Phones that have Qualcomm chips also have an Emergency Download Mode (EDL). Many phones use Qualcomm chips. Getting to EDL mode is relatively easy. Usually booting to edl with adb will work.

adb reboot edl

However, sometimes it continues to reboot normally after about 15 seconds or so. This is because EDL was designed for fixing bricked phones. You must start some tool, such as the MSMDownload as soon as it gets to EDL mode, or it will just try to reboot. MSMDownload is a free download from the internet. You can see the tool in Figure 4.9.

OPERATING SYSTEM DETAILS

The first sections of this chapter provided a general overview of the Android system and introduced you to some tools and techniques. The ADB tool is the most

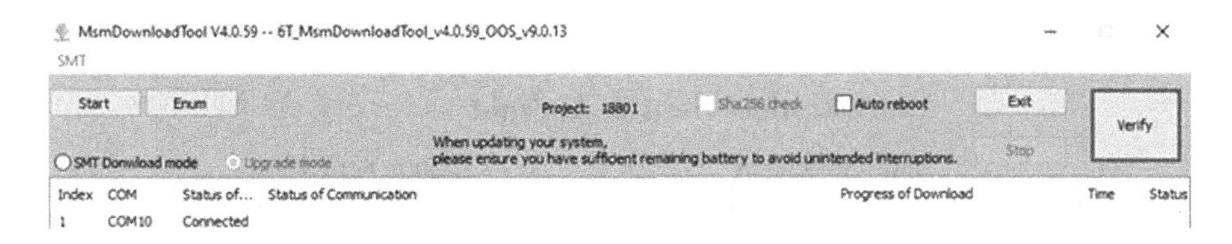

FIGURE 4.9 MSM download.

important. You will use that later when performing Android forensics. In this section you will learn more details of the Android phone.

BOOT PROCESS

Every operating system has a boot process. This begins when the power is turned on and completes when the device's operating system is fully loaded and functioning. Before the device is powered on, the device CPU will be in a state where no initializations have been done. Once the Android device is powered on, execution starts with the boot ROM code. This boot ROM code is specific to the CPU the device is using. This phase includes two steps, A and B:

Step A: When the boot ROM code is executed, it initializes the device hardware and attempts to detect the boot media. Therefore, the boot ROM code scans until it finds the boot media.

Step B: Once the boot sequence is established, the initial boot loader is copied to the internal RAM. After this, the execution shifts to the code loaded into RAM.

This brings us to what is the bootloader? The bootloader is a relatively small program that is executed before the operating system starts to function. Bootloaders are present in all computing devices and used to get the system booted to the point that the operating system can be loaded. In the Android boot loader, there are two stages – Initial Program Load (IPL) and Second Program Load (SPL). As shown in the following screenshot, this involves the three steps explained here:

- Step A: IPL deals with detecting and setting up the external RAM.
- Step B: Once the external RAM is available, SPL is copied into the RAM and execution is transferred to it. SPL is responsible for loading the Android operating system.
- Step C: SPL tries to look for the Linux kernel. It will load this from boot media and will copy it to the RAM. Once the boot loader is done with this process, it transfers the execution to the kernel.

The kernel is the heart of any operating system. And for Android phones that is the Linux kernel. The Linux kernel is the heart of the Android operating system and is responsible for process management, memory management, and enforcing security

TABLE 4.2
Linux Kernels Used in Android

Android Version	Linux Kernel
1.0 to 3.0	used 2.6.xx
4.0 to 6.0	used 3.xx
7.0 to 9.0	used 4.x
10.0	used 4.x
11.0	4.14 to 4.19

on the device. Different versions of Android use different versions of the Linux kernel. The Android kernel is a variation of the Linux kernel, but is quite similar to the standard Linux kernel. This is summarized in Table 4.2.

It should be noted that Android includes a number of different manufacturers. Each manufacturer may alter the Android code base, and even alter the Linux kernel used. So, Table 4.2 is just a general guideline to the kernel version used in specific Android versions.

After the kernel is loaded, it mounts the root filesystem (rootfs) and provides access to system and user data:

Step A: When the memory management units and caches have been initialized, the system can use virtual memory and launch user space processes.

Step B: The kernel will look in the rootfs for the init process and launch it as the initial user space process:

The process rootfs is the place to start and stop searching the doubly linked list of mount points. It is a special instance of ramfs which is a simple file system used to get the system booted. The basic initramfs is the root filesystem image used for booting the kernel provided as a compressed archive.

- If an uncompressed cpio archive exists at the start of the initramfs, extract and load the microcode from it to CPU.
- If an uncompressed cpio archive exists at the start of the initramfs, skip that and set the rest of file as the basic initramfs. Otherwise, treat the whole initramfs as the basic initramfs.
- Unpack the basic initramfs by treating it as compressed (currently gzipped) cpio archive into a RAM-based disk.
- Mount and use the RAM-based disk as the initial root filesystem.

FILE SYSTEMS

File systems are how files are organized on storage devices. Whether that is a traditional hard drive, or a solid-state drive. In chapter 3 we discussed the APFS file system used with iOS devices. There are several file systems used with Android. The major file systems are listed here:

- F2FS: Flash-Friendly File System
- JFFS2: Journaling Flash File System version 2
- YAFFS: Yet Another Flash File System

F2FS: Flash-Friendly File System

The key data structure is the "node". Similar to traditional file structures, F2FS has three types of nodes: inode, direct node, indirect node. F2FS assigns 4 KB to an inode block which contains 923 data block indices, two direct node pointers, two indirect node pointers, and one double indirect node pointer as described below. A direct node block contains 1018 data block indices, and an indirect node block contains 1018 node block indices. The different nodes are shown in Table 4.3.

F2FS divides the volume into six regions. The first is the superblock (SB) which is at the beginning of the partition. There is a second copy of the superblock so that if the first becomes corrupt, the system can recover. Then there is the checkpoint (CP). This contains information about the system including entries of active segments and lists of any orphaned inodes. Next comes the segment information table (SIT) which has a block count and bitmap of the main area blocks. This is followed by the node address table (NAT) which contains the addresses for the nodes. That is in turn followed by the segment summary area (SSA) which contains information about who owns the node blocks. These first five segments are essentially metadata about the volume. They are followed by the main area which has the file and directory data.

JFFS2: Journaling Flash File System version 2

F2FS was originally developed by Samsung. Was first included in the Linux kernel in September 2001. This file system supports two types of nodes: inodes, and dirent nodes. Dirent nodes are directory entries. As the name suggest JFFS2 is a journaling file system. Journaling is a process whereby changes to the file system are logged,

TABLE 4.3
F2FS Nodes

Item	Description
hash	hash value of the file name
ino	inode number
len	the length of file name
type	file type such as directory, symlink, etc

so that in the event of a system failure, the change can be rolled back or recreated. JFFS2 is the default flash file system for the AOSP (Android Open-Source Project) kernels, since Ice Cream Sandwich. JFFS2 is a replacement to the original JFFS.

JFFS2 includes garbage collection. This is a process that fines file system blocks that were not released by a process but should have been. The garbage collection then releases them. This is similar to memory garbage collection that is found in Java and .Net programming languages.

YAFFS: Yet Another Flash File System

With YAFFS, data is written as an entire page that includes the file metadata as well as the data. In YAFFS terminology a page is referred to as a chunk. Each new file is given a unique object ID number so that it can be identified. YAFFS uses a tree data structure of the physical location of the chunks/pages in the system. Version 2 was the default AOSP flash file system for kernel version 2.6.32. YAFFS2 is not supported in the newer kernel versions and does not appear in the source tree for the latest kernel versions from kernel.org. However, individual mobile device vendors may continue to support YAFFS2.

SECURITY

As Android is marketed on a wide range of devices, made by different manufacturers, some of the Android security features are unique to specific manufacturers and models. Cryptography is a critical part of security with any mobile device.

In February 2019, Google unveiled Adiantum, an encryption cipher designed primarily for use on devices that do not have hardware-accelerated support for the Advanced Encryption Standard (AES), such as low-end devices. Adiantum is a cipher construction for disk encryption, which uses the ChaCha cipher and Advanced Encryption Standard (AES) ciphers, and Poly1305 cryptographic message authentication code (MAC). ChaCha is a variant of the Salsa stream cipher.

You are probably at least somewhat familiar with AES but may not be familiar with Salsa or ChaCha. In 2013, Mouha and Preneel published a proof that 15 rounds of Salsa20 was 128-bit secure against differential cryptanalysis.[1] Specifically, Salsa has no differential characteristic with higher probability than 2^{130}, so differential cryptanalysis would be more difficult than 128-bit key exhaustion. This is one example of studies done with the Salsa cipher that show it to be a robust cipher.

Aside from cryptography, the Android operating system also supports application security., Android uses the Linux user-based protection model to isolate applications from each other. In Linux systems, each user is assigned a unique User ID (UID) and users are segregated so that one user does not have access to the data of another. All resources under a particular user are run with the same privileges. In the same manner, each Android application is assigned a UID and is run as a separate process. This effectively puts each Android app in an isolated sandbox.

SELinux (Security-Enhanced Linux) is also a security enhancement available for Android devices. This began with Android version 4.2. Android uses SELinux to impose mandatory access control that ensures applications work in isolated environments. Therefore, even if a user installs a malicious app, the malware cannot easily access the operating system or other applications.

GENERAL ARCHITECTURE

While Android includes a number of different vendors, each with their own modifications, the general overview of the Android architecture is common to all Android devices. This common architecture is shown in Figure 4.10.

The first item of note in Figure 4.10 is the Android runtime (ART). This is the application environment used by Android. It is a virtual machine. Each application is executed in its own copy of the ART. ART replaces the earlier Dalvik system. Dalvik used just in time compilation. That means that an app was compiled to executable code when it was executed. This saved storage space but slowed app launching. ART uses ahead of time compiling (AOT). As soon as an app is installed, it is compiled into executable code. This speeds up performance, but with a cost so storage. However, modern devices have ever increasing storage space.

TOOLS

There are various tools made for Android phones. Some are for specific manufacturers, and others work with a range of Android phones. Earlier in this chapter MSDownload was mentioned. That was one example of an Android specific tool. It happens that most of these tools are open source and free downloads. Those that do have a cost, have a minimal cost.

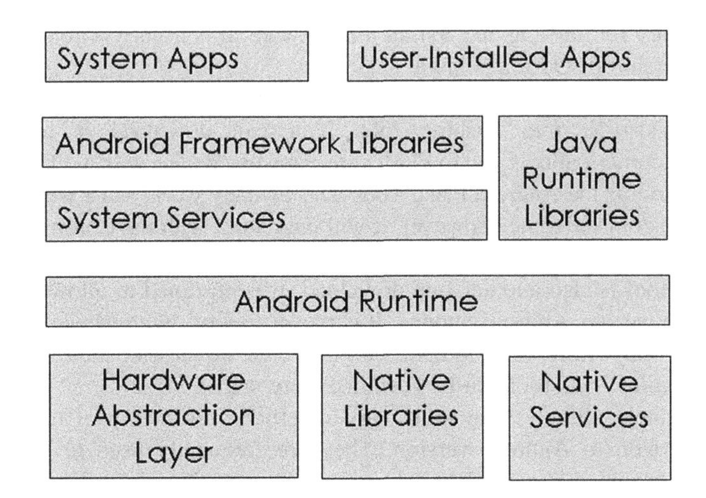

FIGURE 4.10 Android architecture.

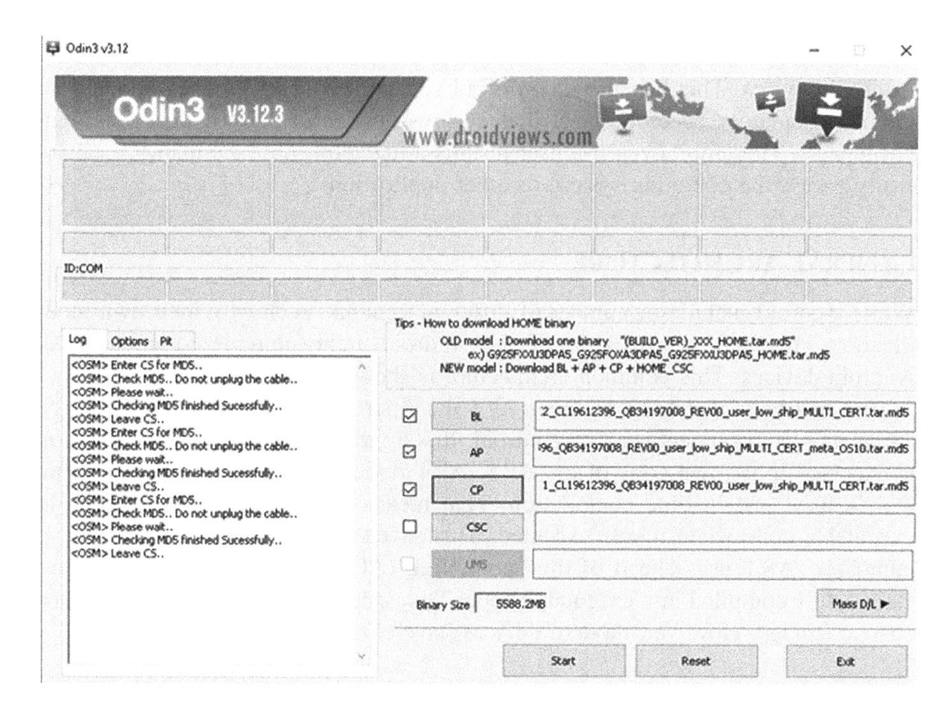

FIGURE 4.11 Odin.

Another tool is Odin3. This tool is explicitly for Samsung phones. You can see the tool in Figure 4.11.

Odin can be downloaded from https://samsungodin.com/. This tool is used to flash images onto a Samsung phone. This includes stock or custom firmware images. And, at least in theory, root packages that will root the Samsung phone. The tool was created for internal use by Samsung in their authorized service centers, and in their factory for flashing and testing the software. It is usually a good idea to use the latest version of Odin available.

Another Samsung only software is SamFirm. This is shown in Figure 4.12.

This is also a free flashing tool. You can download it from https://samfirmtool.com/. It allows you to Flash your Samsung device with a different image.

Another tool is the Android Flash Tool. This actually works via a website, https://flash.android.com/welcome. However, it will only work with the Chrome browser. It has the advantage of being quite simple to use. The tool is shown in Figure 4.13.

SPFlash tool is also another free download. It is designed to allow you to flash images onto various Android phones. It is meant to work with MediaTek Android phones but may work with others. SPFlash can be downloaded from https://spflashtool.com/. This tool can be seen in Figure 4.14.

Unfortunately, all of these tools vary in efficacy from model to model and Android version to Android version. They are frequently used to help recover bricked phones or to bypass phone security. However, the use of these tools is not without some risk. It is possible to completely brick your phone using these tools.

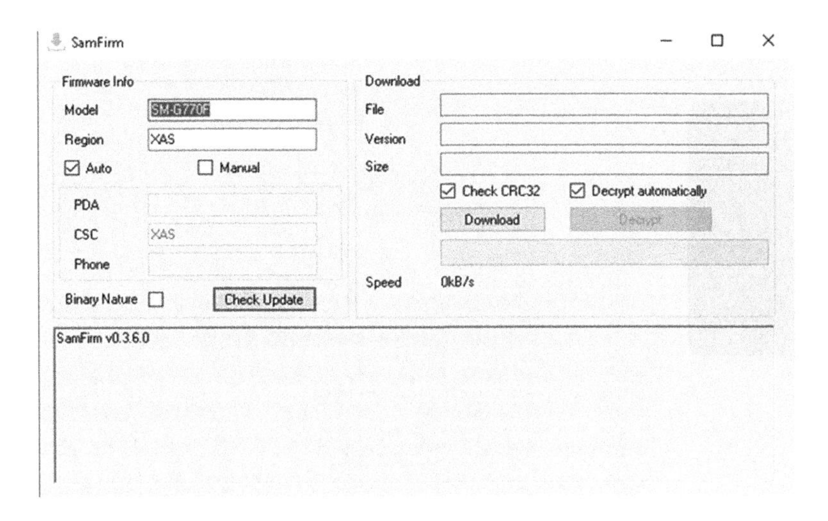

FIGURE 4.12 SamFirm.

Welcome to Android Flash Tool

You can use this tool to install Android builds on your devices.

This tool allows you to flash Android onto recent Pixel phones and some Android development devices (view full list) and requires 10GB of available storage on your computer. This tool doesn't support flashing Android onto tablets or Chrome OS devices.

Install an Android build in three easy steps

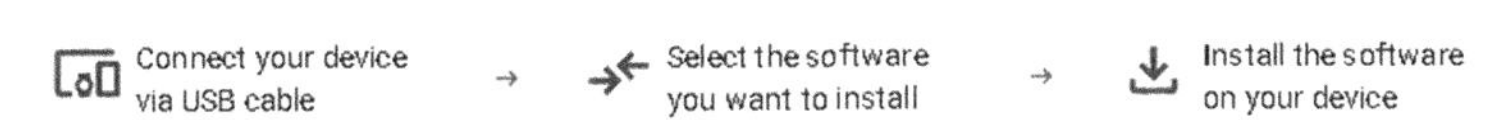

FIGURE 4.13 Android flash tool.

DEEP INTO ANDROID CODE

As you probably know, Android is an open-source operating system. That means the source code is freely available. Most vendors of commercial products alter the public code somewhat, but the public source code is going to be mostly the same as what runs on your phone, regardless of the vendor. The website https://cs.android.com/ allows you to view the code online. You can search the code for specific items, or peruse the file structure, as shown in Figure 4.15.

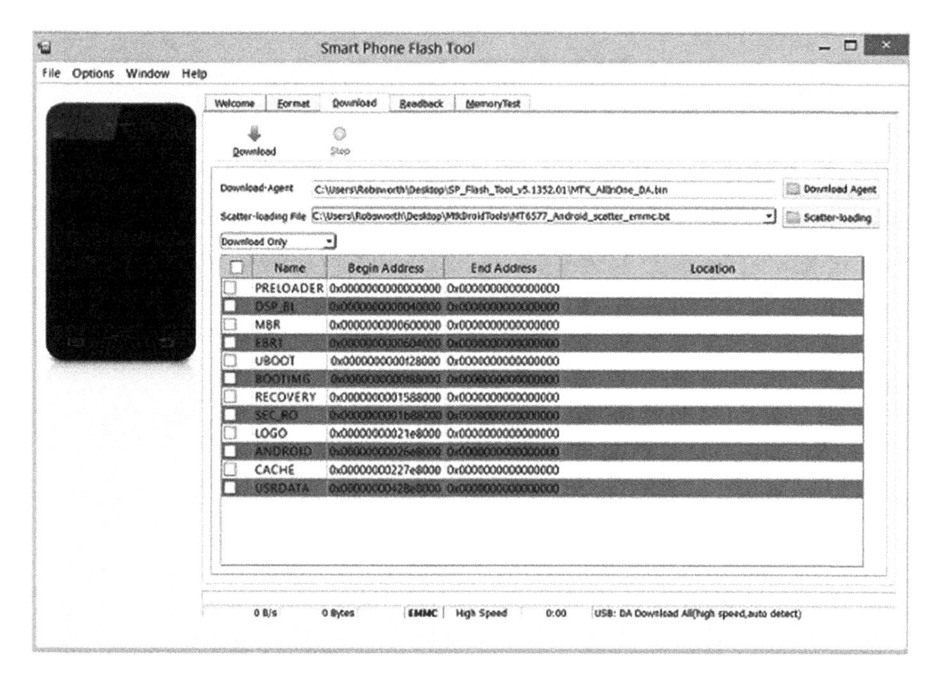

FIGURE 4.14 Android flash tool.

It should be obvious, that this is most useful to those readers with a strong programming background. But if you truly wish to understand Android devices at a deep level, it is certainly worthwhile to become familiar with the source code. Figure 4.16, as an example, shows the function placeCall which is central in actually placing outgoing calls.

If you do have a programming background, particularly in C++ and Java, you are encouraged to at least spend some time perusing the Android code and understanding the general flow. This can be useful in understanding why the phone works in the way it does. It can also be useful for searching out vulnerabilities that may exist in Android phones.

CHAPTER SUMMARY

In this chapter, the Android operating system was explored in some detail. The general understanding of the operating system and file systems will be important as you proceed through the rest of this book. Specifically, in chapters six and nine, you will apply the knowledge from this chapter. The ADB tool is of particular importance. It is important that you be familiar with this tool.

The knowledge of the boot system, file systems, and security provides a general context for Android forensics. The more you understand about the device you wish to examine, the better. The security features are also important. You may have need to circumvent security to extract data from a suspects phone. It is important to understand what is possible, and what is highly unlikely.

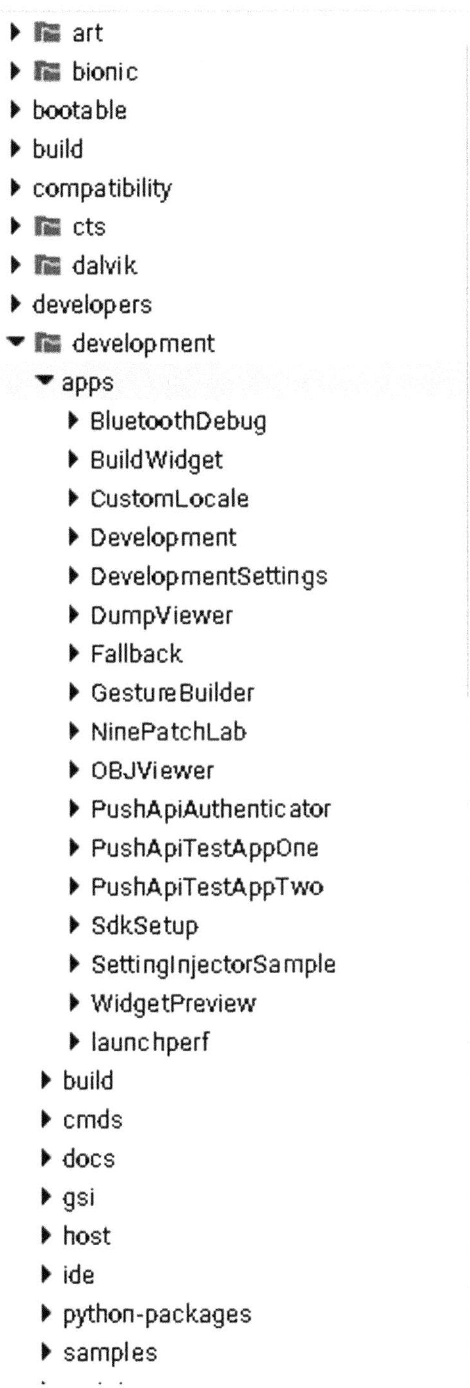

- art
- bionic
- bootable
- build
- compatibility
- cts
- dalvik
- developers
- development
  - apps
    - BluetoothDebug
    - BuildWidget
    - CustomLocale
    - Development
    - DevelopmentSettings
    - DumpViewer
    - Fallback
    - GestureBuilder
    - NinePatchLab
    - OBJViewer
    - PushApiAuthenticator
    - PushApiTestAppOne
    - PushApiTestAppTwo
    - SdkSetup
    - SettingInjectorSample
    - WidgetPreview
    - launchperf
  - build
  - cmds
  - docs
  - gsi
  - host
  - ide
  - python-packages
  - samples

FIGURE 4.15 Android source code file structure.

```
/**
 * @see android.telecom.TelecomManager#placeCall
 */
@Override
public void placeCall(Uri handle, Bundle extras, String callingPackage,
        String callingFeatureId) {
    try {
        Log.startSession("TSI.pC");
        enforceCallingPackage(callingPackage);

        PhoneAccountHandle phoneAccountHandle = null;
        if (extras != null) {
            phoneAccountHandle = extras.getParcelable(
                    TelecomManager.EXTRA_PHONE_ACCOUNT_HANDLE);
            if (extras.containsKey(TelecomManager.EXTRA_IS_HANDOVER)) {
                // This extra is for Telecom use only so should never be passed in.
                extras.remove(TelecomManager.EXTRA_IS_HANDOVER);
            }
        }
        boolean isSelfManaged = phoneAccountHandle != null &&
                isSelfManagedConnectionService(phoneAccountHandle);
        if (isSelfManaged) {
            mContext.enforceCallingOrSelfPermission(Manifest.permission.MANAGE_OWN_CALLS,
                    "Self-managed ConnectionServices require MANAGE_OWN_CALLS permission.");

            if (!callingPackage.equals(
                    phoneAccountHandle.getComponentName().getPackageName())
                    && !canCallPhone(callingPackage, callingFeatureId,
                    "CALL_PHONE permission required to place calls.")) {
                // The caller is not allowed to place calls, so we want to ensure that it
                // can only place calls through itself.
                throw new SecurityException("Self-managed ConnectionServices can only "
                        + "place calls through their own ConnectionService.");
            }
```

FIGURE 4.16 Android source code placeCall function.

CHAPTER 4 ASSESSMENT

1. _______ is the default flash file system for the AOSP (Android Open-Source Project)
    a. APFS
    b. YAFSS
    c. F2FS
    d. JFFS2
2. Which directory should you look in for frequently accessed data?
    a. bin
    b. data
    c. cache
    d. sbin
3. What ADB command will list all the Linux commands supported by the device?
    a. adb shell Linux
    b. adb shell Command
    c. adb shell ls system/bin
    d. adb shell ls commands

4. What does the command *pm list packages -e* do?
   a. List only enabled packages.
   b. List only executable packages
   c. List only excluded packages.
   d. List only extraneous packages
5. What command will tell you if the device is OEM unlocked?
   a. adb shell oem device-info
   b. fastboot oem device-info
   c. adb shell device-info
   d. fastboot device-info
6. What is the Initial Program Load responsible for?
   a. loading system apps
   b. loading the first user apps
   c. setting up the external RAM
   d. setting up ART

CHAPTER 4 LABS

For these labs you will need an Android phone. Even an older model, or one with defects such as a broken screen will be adequate. As long as it powers on, that is the issue. Such phones can be found used on E-Bay, Amazon.com, and phone repair stores for approximately $20 US Dollars.

LAB 4.1 CODES

Try as many of the various "secret codes" as you can. Not all work on all models, so don't expect to succeed with all commands.

##4636#*#* Testing Menu
##4636#*#* Display Info about device
##34971539#*#* Camera Information
##273283*255*663282*#*#* Backup all media files
##232339#*#* OR *#*#526#*#* Wireless LAN Test
Change Power button behavior *#*#7594#*#*
##1472365#*#* Quick GPS Test
##197328640#*#* Test mode for service activity
##232338#*#* Wi-Fi Mac Address
##1575#*#* Another GPS test
##0842#*#* Vibration and Backlight test
##2663#*#* Check touch screen version
##2664#*#* Touch Screen test
##0588#*#* Proximity sensor test
##3264#*#* RAM version
##232331#*#* Bluetooth test
*#0228# Battery Status

##44336#*#* Shows Build time change list number.
*#0589# Light sensor test
*#7353#Quick test menu
*#99#...........Toggle always-on display on or off
*#800#..........Log test settings
*#801#..........Engineering switch test mode
*#802#..........GPS TTFF (Time-To-First-Fix) test mode
*#803#..........Engineering Wi-Fi setting
*#804#..........Automatic disconnect test mode.
*#805#..........Engineering bluetooth test mode
*#806#..........Engineering aging test mode
*#807#..........Engineering automatic test mode
*#809#..........Enter engineering mic echo test mode.
*#814#..........Automatically searches for available TDSCDMA carriers.
*#824#..........Automatically searches for available WCDMA carriers.
*#834#..........Automatically searches for available LTE carriers.
*#844#..........Automatically searches for available GSM carriers.
*#900#..........Test photograph RGB (Red, Green, & Blue tint)
##0*#*#**.....LCD display test
##0283#*#*....Packet loopback
##0289#*#*....Melody test
##0588#*#*....Proximity sensor test
##0673#*#*....Melody test
##0842#*#*....Test for vibration and backLight functionality
##1575#*#*....Advanced GPS testing
##2664#*#*....Touch screen test
##7668#*#*....Checks for root
##232331#*#* - Bluetooth test

Lab 4.2 ADB

First backup your phone using ADB

```
adb backup -all -f C: \backup.ab
```

Next connect to the device starting with

```
adb devices
```

You may need to troubleshoot issues here.

Then try adb shell

This will be followed by navigating the phones file directory using:

```
ls
```

```
ls -f
```

Then try the other commands we discussed:

```
ps
netstat
dumpstate
getprop ro.product.model
getprop ro.build.version.release
getprop ro.serialno
adb devices -l
adb get-state
adb shell list packages
adb reboot fastboot (for this one just look at the menu options then reboot normal)
```

Try as many shell commands as you can. Get quite familiar with the ADB shell.

Lab 4.3

Try to OEM unlock the phone as per instructions in this chapter. This may not work for phones, particularly if they are network locked.

NOTE

1 Mouha, N., & Preneel, B. (2013). A Proof that the ARX Cipher Salsa20 is Secure against Differential Cryptanalysis. *IACR Cryptol. ePrint Arch., 2013*, 328.

Section II

Forensic Techniques

5 iOS Forensics

The process of iOS forensics includes all the standard mobile forensics as well as general forensics principles. However, it also has specialized techniques. This includes a number of standards such as NIST as well as tools. Several tools are covered in this chapter. It is critical to understand the techniques as well as the tool. However, one must keep in mind the general forensics principles and mobile device concepts that have been discussed previously in this book.
Regardless of the manufacturer, or model of a mobile device, a treasure trove of evidence can frequently be found on such devices. This introductory section will be essentially the same as what will be given in chapter 6, due to the fact that the issues discussed here are common to all cell phones and other mobile devices.

One reason why mobile devices provide so much evidence is the ubiquitous nature of such devices. Many people would not think of going anywhere without their smartphone. Others carry a tablet with them everywhere. More and more of our lives are conducted on our devices. From ordering food to communicating with friends. Because of the pervasive nature of mobile devices, mobile forensics is important in all types of investigations.

Items you should attempt to recover from a mobile device include the following:

- Call history
- Emails, texts, and/or other messages
- Photos and video
- Phone information
- Global positioning system (GPS) information
- Network information

The call history lets you know who the user has spoken to and for how long. This information is often important to all types of investigations. Of course, this information is easily erasable, but many users don't erase their call history. Or perhaps the suspect intended to delete this data and simply did not get to it yet. Usually call history does not provide direct evidence of a specific crime, so much as it provides supporting evidence and general intelligence about the suspect's activity.

However, in a cyberstalking case, the call history can be of central importance. The call history can show a pattern of contact with the victim. However, in other cases, it provides only circumstantial evidence. For example, if John Smith is suspected of drug dealing and his call history shows a pattern of regular calls to a known drug supplier, by itself this is not adequate evidence of any crime. However, it aids the investigators in getting an accurate picture of the entire situation.

Smartphones also allow text messages and email. There is also a wide range of chat apps such as Snapchat, Viber, WhatsApp, WeChat, Signal, etc. Many phone

DOI: 10.1201/9781003118718-5

forensics tools will retrieve information from some apps. However, given there are hundreds of apps, no tool could possibly retrieve data from all of them. In fact, most tools only get data from a few dozen. In chapter 8 we will discuss SQLite forensics, which will allow you to extract data from the app databases directly.

Photos and videos can provide direct evidence of a crime. In the case of child pornography cases, the relevance is obvious. But there are certainly other crimes where the evidence is found in photos and videos on the suspect's own device. The breach of the U.S. Capitol on January 6, 2021 is a case in point. Many of these who breached the Capitol building took selfies. In many cases uploading those to social media platforms.

This is in no way anomalous. It may surprise you to find that it is quite common for at least some criminals to have incriminating photos and videos on their devices. There are cases of gang members with photos depicting them with illegal firearms and narcotics, thieves videoing themselves during a theft, and many more similar examples. It is always worthwhile to examine photos and videos on a phone.

Information about the phone should be one of the first things you document in your investigation. This will include model number, IMEI number, serial number of the SIM card, operating system, and other similar information. The more detailed, descriptive information you can document, the better. It is important to fully document the details of the phone.

Global positioning system information has become increasingly important in a variety of cases. GPS information, even if it is not exact. can determine if a suspect was in a particular area at the time of the crime. That data can also provide an alibi. If a crime is committed, and phone GPS indicates the person suspected of the crime was actually 100 miles away at the time, this is a compelling alibi. GPS information has begun to play a significant role in contentious divorces, for instance. If someone suspects a spouse of being unfaithful, determining that the spouse's phone and his or her car were at a specific motel when he or she claimed to be at work can be important.

It should be noted that with older phones, the phone did not have true GPS. Rather than use GPS satellites to determine location (which is the most reliable method), they instead would use triangulation of signal strength with various cell towers. This could lead to inaccuracies of up to 50 to 100 feet. However, this situation has changed. Most modern phones and/or the apps on the phone use true GPS for much more accurate data.

The use of Wi-Fi along with GPS will improve accuracy of GPS. The reason for this is that various organizations, including Google, track the Basic Service Set Identifier (BSSID) used by wireless routers, and correlate it with physical addresses. The BSSID is a unique address that identifies the access point/router that creates the wireless network. To identify access points and their clients, the access points MAC address is used. This implies that if your phone connects to a wireless access point, then even with no other data, the phone's location can be pinpointed to within a reasonably close distance of that access point. Network information is also important. Many phones will store the various Wi-Fi hotspots they have connected to. This data could give you an indication of where the phone has been. For example, if the phone belongs to someone suspected of stalking a victim, and the suspect's phone network records show he or she has frequently been using Wi-Fi networks in close proximity to the victim's home, this can be important evidence.

Regardless of the make or model of a mobile phone, these mobile devices are a repository of invaluable data. As was stated earlier in this chapter, so much of our lives is tied to our mobile devices.

FORENSIC PROCEDURES

There are various standards the provide guidelines for forensic examinations. Mobile forensics is no exception to the need for standards. One standard procedure that you should absolutely follow is to put the phone in airplane mode while working with it. You do not want anyone to be able to remotely access the device. It is also important that you make as few changes on any device you are examining, and airplane mode will assist you with that.

One group that is at the forefront of digital forensics procedures is the Scientific Working Group on Digital Evidence (https://www.swgde.org/). SWGDE provides guidance on many digital forensics topics. Related to mobile device forensics, SWGDE provides a general overview of the types of phone forensic investigations:

Mobile Forensics Pyramid – The level of extraction and analysis required depends on the request and the specifics of the investigation. Higher levels require a more comprehensive examination, additional skills and may not be applicable or possible for every phone or situation. Each level of the Mobile Forensics Pyramid has its own corresponding skill set. The levels are:

1. Manual – A process that involves the manual operation of the keypad and handset.
   display to document data present in the phone's internal memory.
2. Logical – A process that extracts a portion of the file system.
3. File System – A process that provides access to the file system.
4. Physical (Non-Invasive) – A process that provides physical acquisition of a phone's data without requiring opening the case of the phone.
5. Physical (Invasive) – A process that provides physical acquisition of a phone's data requiring disassembly of the phone providing access to the circuit board (e.g., JTAG).
6. Chip-Off – A process that involves the removal and reading of a memory chip to conduct analysis.
7. MicroRead – A process that involves the use of a high-power microscope to provide a physical view of memory cells.

Regardless of the tool you choose, you should ensure that the examination chosen is adequate for your forensic needs. Levels 5 and 6, JTAG and MicroRead, are not as common as the other methods. While level 1, manual, is usually not detailed enough. Level 4 is the most common used, but sometimes a level 2 or 3 can be sufficient.

The United States National Institute of Standards (NIST) published NIST SP 800-101 *Guidelines on Mobile Device Forensics*. This document was mentioned in chapter 2. It has a wealth of guidance. It discusses in some detail mobile device hardware, and specific operating system issues (iOS and Android). Furthermore, it describes the types of data to look for when extracting evidence. That list includes date/time settings, language settings, contacts, messages, call logs, photos, videos, audio, web browser activity, geolocation data, and app-related data.

It is also important to have the appropriate tools for the examination. Fortunately, NIST also provides guidance on this issue. The NIST-sponsored CFTT – Computer Forensics Tool Testing Program (http://www.cftt.nist.gov/) provides a measure of assurance that the tools used in the investigations of computer-related crimes produce valid results. Testing includes a set of core requirements as well as optional requirements. It is a good idea to refer to these standards when selecting a tool.

NIST also provides general guidelines on how to write a report for a mobile device forensic report. The guidelines are of what to include are:

- Descriptive list of items submitted for examination, including serial number, make, and model
- Identity and signature of the examiner
- The equipment and setup used in the examination
- Brief description of steps taken during examination, such as string searches, graphics image searches, and recovering erased files
- Supporting materials, such as printouts of particular items of evidence, digital copies of evidence, and chain of custody documentation
- Details of findings:
    - Specific files related to the request
    - Other files, including deleted files, that support the findings
    - String searches, keyword searches, and text string searches
    - Internet-related evidence, such as website traffic analysis, chat logs, cache files, email, and news group activity
    - Graphic image analysis
    - Indicators of ownership, which could include program registration data.
    - Data analysis
    - Description of relevant programs on the examined items.
    - Techniques used to hide or mask data, such as encryption, steganography, hidden attributes, hidden partitions, and file name anomalies
- Report conclusions

The United States Department of Justice[1] states

> The examiner is responsible for completely and accurately reporting his or her findings and the results of the analysis of the digital evidence examination. Documentation is an ongoing process throughout the examination. It is important to accurately record the steps taken during the digital evidence examination.

The well-respected SANS institute publishes guidelines[2] on digital forensics reports. They state a digital forensics report should include overview/case summary, details on the forensic acquisition and exam, details of all steps taken, etc. The Reference Manual on Scientific Evidence: Third Edition[3] states “Under Federal

Rule of Civil Procedure 26(a)(2)(B)(i), the expert report must contain the basis and reasons for all opinions expressed, and certainly the expectation is that oral testimony will do the same." An inadequate report can lead to the report and the associated testimony being excluded at trial.

The important thing to remember when creating any forensics report is that it is quite difficult to put too much detail in. It is much better to have information that is ultimately not necessary, than to wish you had captured data later. When forensics cases go to court, there may be an opposing forensic expert. The idea of a forensic report is that it is so detailed, that any competent forensic examiner can recreate the steps you took and either verify or refute your conclusions. This means that your report should quite literally create a roadmap of your investigation, and one that any competent examiner can follow.

IOS SPECIFIC PROCESSES

While there are general rules that apply to any mobile device, there are also rules that are specific to iOS devices. This is particularly important when using a Windows PC to analyze an iOS device. You need to ensure that the PC does not write any data to the iOS device. Once you are ready to seize evidence from the mobile device, remember the following rules:

- If you are going to plug the phone into a computer, make sure the phone does not synchronize with the computer. This is particularly important with the iPhone, which routinely auto-syncs.
- Follow the same advice you follow for PCs. Make sure you touch the evidence as little as possible, and document what you do to the device.

One of the most important things to do is to make sure you don't accidentally write data to the mobile device. For example, if you plug an iPhone into your forensic workstation, you want to make sure you don't accidentally write information from your workstation to the iPhone.

If the forensic workstation is a Windows machine, you can use the Windows Registry to prevent the workstation from writing to the mobile device. Before connecting to a Windows machine, find the Registry key (HKEY_LOCAL_MACHINE \System\CurrentControlset\StorageDevicePolicies) and set the value to 0×00000001, then restart the computer. This prevents that computer from writing to mobile devices that are connected to it.

TOOLS

Beyond forensic techniques and processes are the specific tools you may use. Each tool has its advocates and detractors. The purpose of this book is not to endorse a particular tool. It is a goal to ensure you have a general understanding of the various tools so that you can make an informed choice.

Cellebrite

This is probably the most widely known phone forensics tool. It is used heavily by federal law enforcement. It is also well respected in the industry. It is a very robust and effective tool. The only downside to Cellebrite that I am aware of is its high cost. It is the most expensive phone forensics tool I am aware of. Most of the Cellebrite literature won't tell you directly the cost, you need to speak to a salesperson to get a bid. In general, one can expect to spend in the neighborhood of $10,000 for a Cellebrite license. Cellebrite is an Israeli company known not only for their tools, but also for mobile forensics research.

While this tool is quite popular with law enforcement, and well respected, there is not a screenshot-by-screenshot description of the tool in this chapter. This is primarily due to the fact the Cellebrite is not a single tool. There are a number of tools available from Cellebrite including[4]:

- Cellebrite UFED
- Cellebrite Physical Analyzer
- Cellebrite UFED Cloud
- Cellebrite Premium
- Cellebrite Blacklight
- Cellebrite Commander

Even these products have variations. For example, UFED has UFED Touch 2, UFED Touch 2 Ruggedized, and others. It would take an entire book to adequately describe the various Cellebrite products. And, as was discussed earlier, each can be rather expensive. It is also the case that the Cellebrite tools usually require formal training, beyond the scope of this chapter or book. The primary focus in this chapter will be on tools that are more affordable. But more importantly, tools that don't require extensive training.

iMazing

The tool iMazing was designed to be a manager for iOS devices. It is quite affordable at $49.99 for a lifetime license for up to 3 devices, or $59.99 per year for unlimited licenses. It can be found at https://imazing.com/. Even though it was not designed for forensics, it does gather much of the data a forensic examiner requires, and with an easy-to-use interface. Furthermore, the tool has been the subject of various scientific articles[5,6] and even forensics books.[7] Figure 5.1 shows the main screen of iMazing.

The tool detects any iPhone connected to the forensic computer and displays device details on the lower right, while giving easy to understand icons in the center and far left, that one can click on to get specific data. Figure 5.2 shows the messages retrieved from the iPhone. For privacy reasons, some of the data is redacted.

You will note the icons on the bottom of the screen that allow the data to be exported to PDF, Excel, CSV, or test files. You can also export any attachments. This tool does not have features such as the ability to attempt to break encryption, recover deleted files, or to plot data on a map for easy geolocation. But the basic data needed in many forensic investigations can be easily retrieved and exploited.

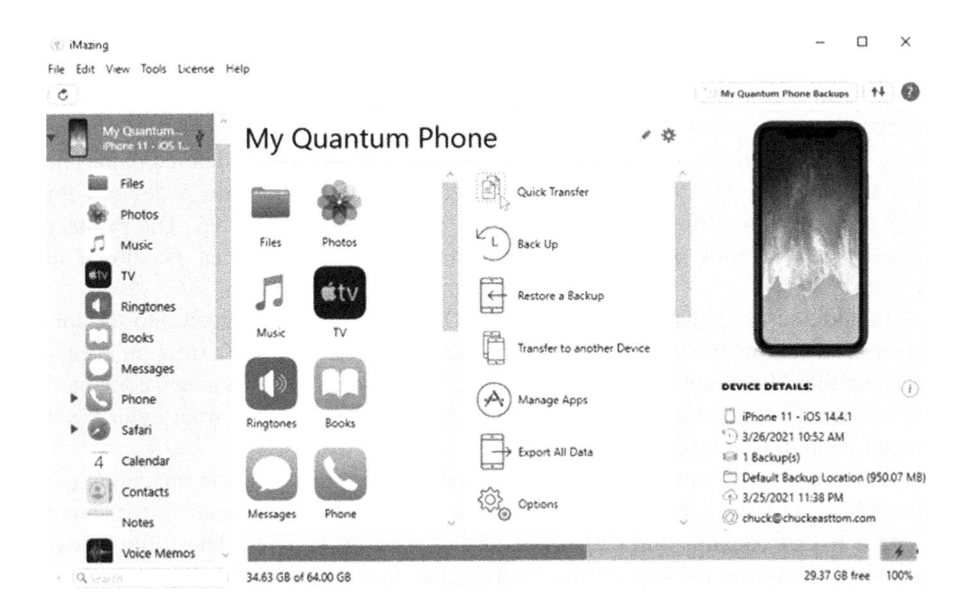

FIGURE 5.1 iMazing initial screen.

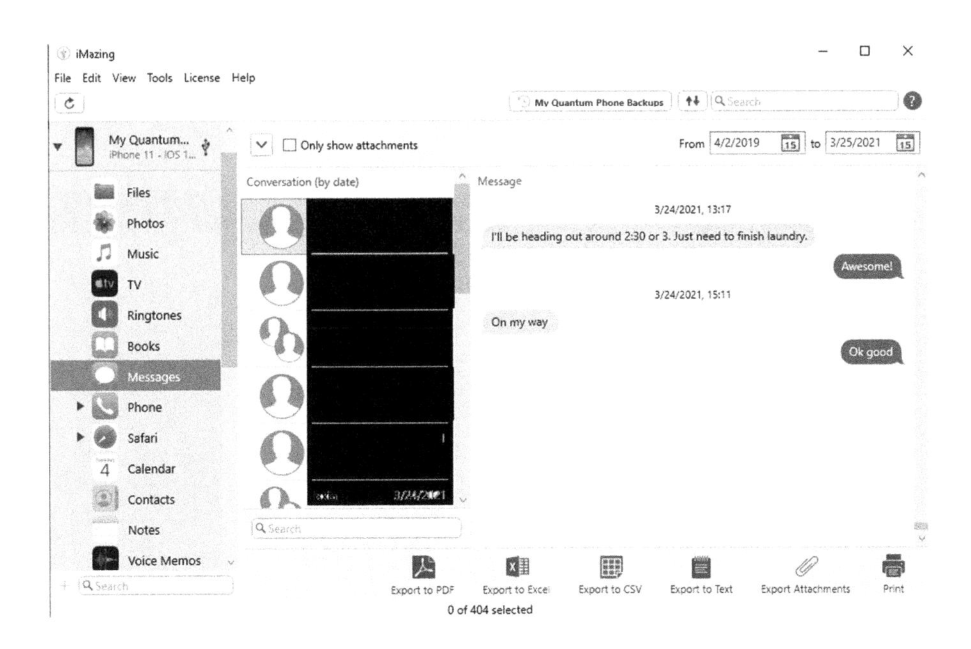

FIGURE 5.2 iMazing messages.

iMyPhone D-Back

This tool is designed specifically to recover information from an iOS device (https://www.imyfone.com/). This makes it a good combination with some of the other tools discussed in this chapter that are not able to recover deleted data. The license is either $49.95 per year or a lifetime license for up to 5 devices for $69.95. There is a lifetime plan for unlimited devices for $299.95. The primary benefit of this tool is its intuitive interface. The main screen is shown in Figure 5.3.

In addition to recovering data from an iOS device that is plugged into the forensic workstation, this tool will allow the examiner to recover data from an iTunes Backup file. Many users backup to the cloud, but if there is a backup on a computer, it is certainly a good idea to examine that backup to determine what evidence it might contain (Figure 5.4).

This tool will locate any backups that are on the forensic machine the iMyPhone D-Back software is running on. The forensic examiner is presented with a list of backups and can select any of them to recover. The forensic examiner selects the backup file of interest and then extracts the data from that backup (Figure 5.5).

w?>The tool presents a set of data similar to what is extracted from a physical device. There is the call history, contacts, messages, and even third-party apps. This is an easy-to-use interface that can be quickly navigated to find data. This is shown in Figure 5.6.

This tool can also examine data that is backedup to the iCloud. This will require the credentials to log in to iCloud. However, that may not be as difficult to find as it

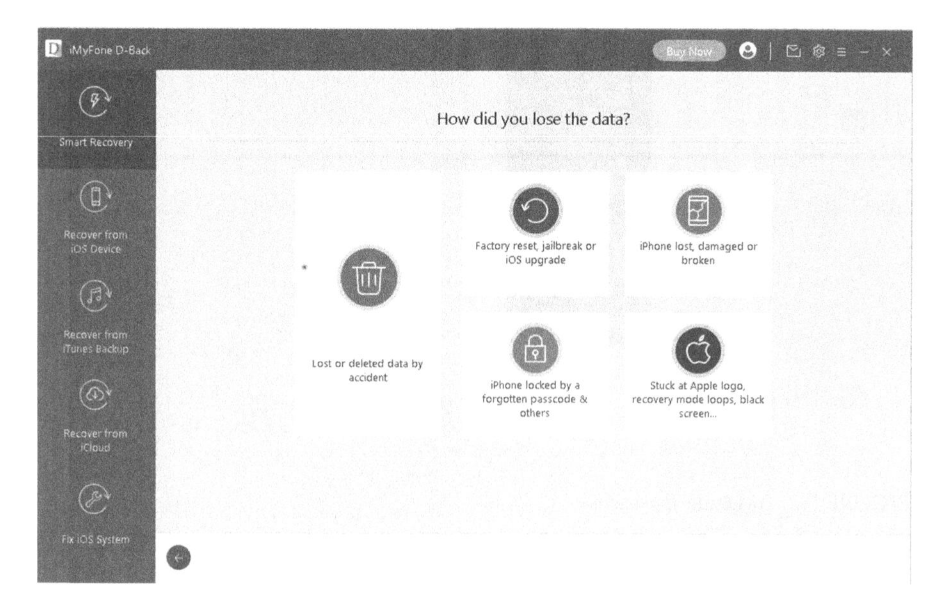

FIGURE 5.3 iMyPhone D-Back main screen.

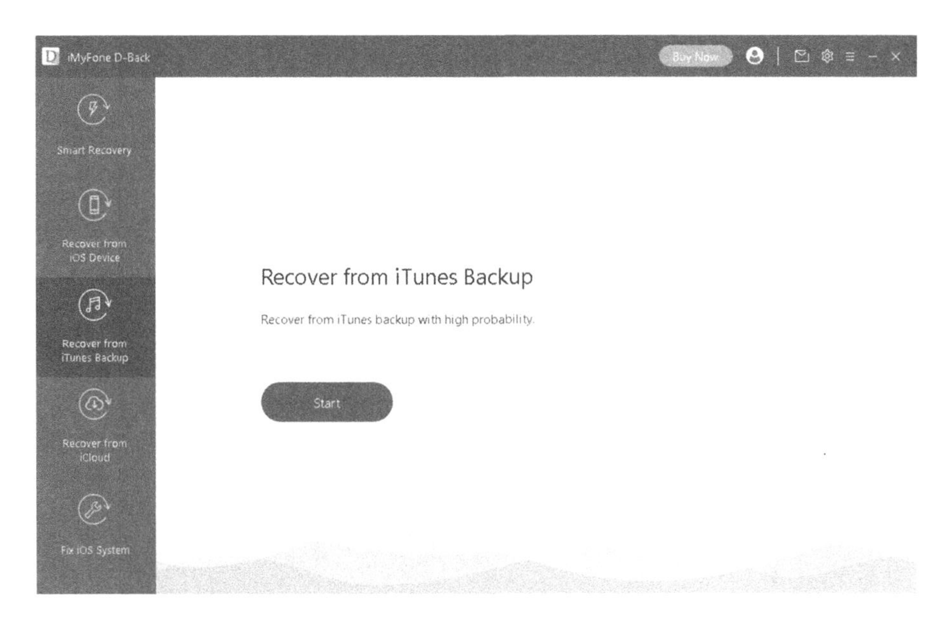

FIGURE 5.4 iMyPhone D-Back recover backup.

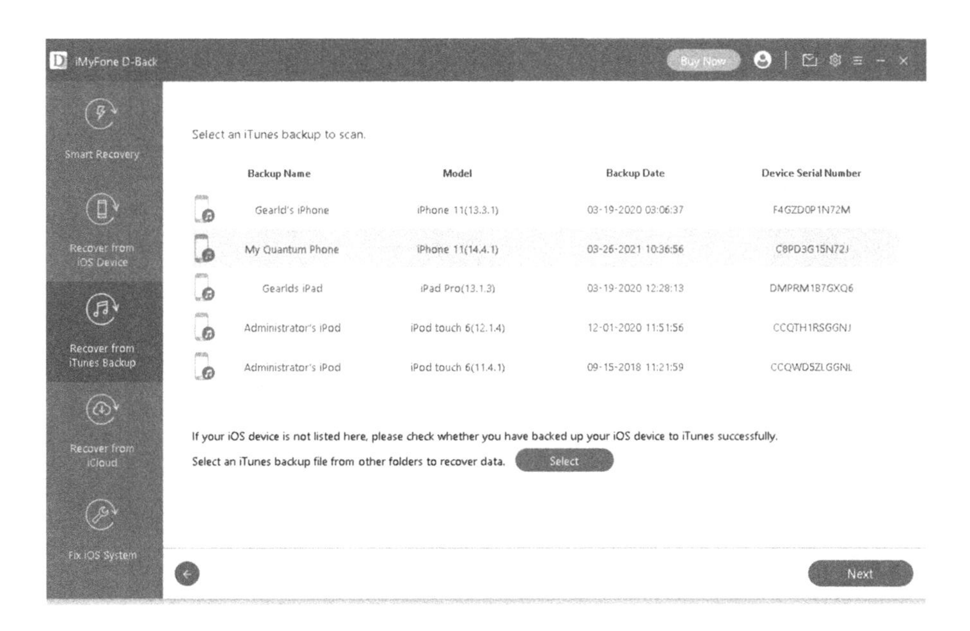

FIGURE 5.5 iMyPhone D-Back list of backups.

might sound. If the suspect uses a Windows computers, those are usually easier to extract information from than phones. Of particular interest will be extracting any stored passwords from the suspect's Windows computer. The iCloud wizard is shown in Figure 5.7.

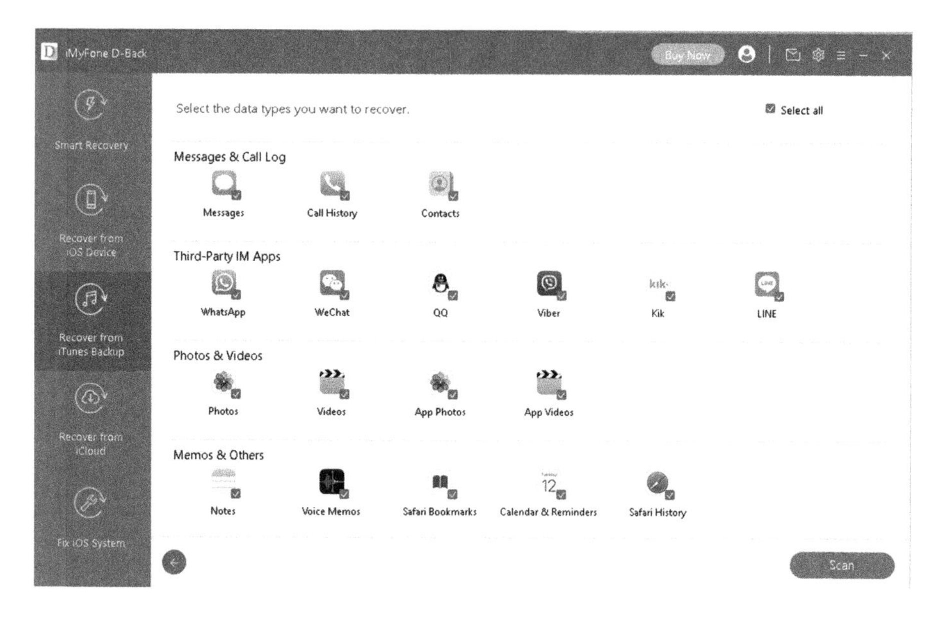

FIGURE 5.6 iMyPhone D-Back data from backup.

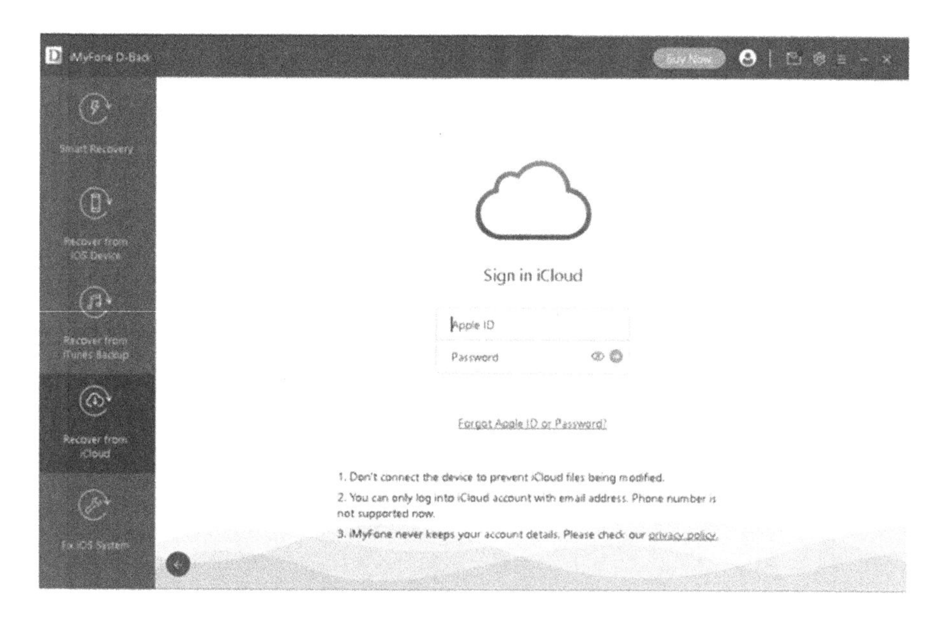

FIGURE 5.7 iMyPhone D-Back iCloud extraction.

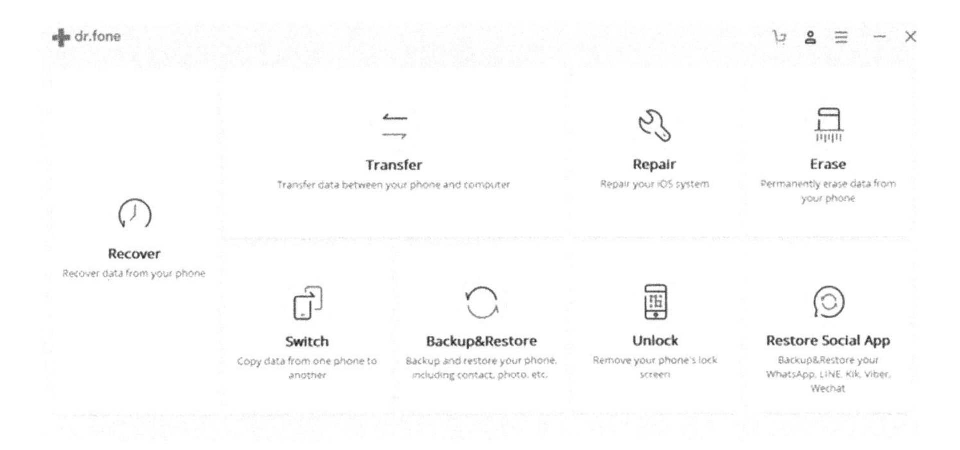

FIGURE 5.8 Dr. Fone Main Screen.

Dr. Fone

Dr. Fone is a widely used tool for mobile device recovery and transferring of data. This makes it particularly interesting for forensics. This tool is very inexpensive and works with both iPhone and Android. The tool can be found at https://drfone.wondershare.net. The full version is $139.95. The main screen is shown in Figure 5.8.

The information tab will allow you to view SMS messages and phone numbers. These are often critical to a digital forensic investigation. Using Dr. Fone you can see the file system discussed earlier. This is shown in Figure 5.9.

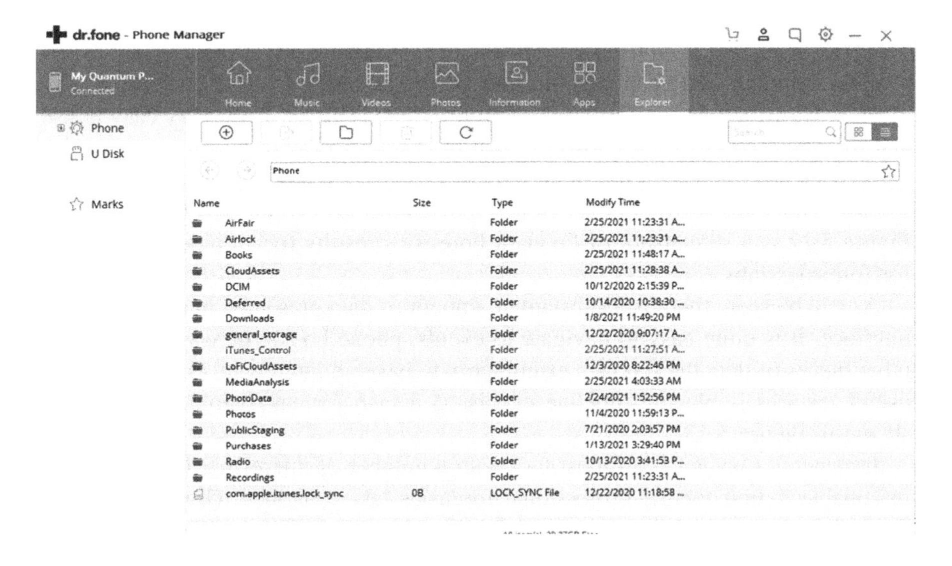

FIGURE 5.9 Dr. Fone file system.

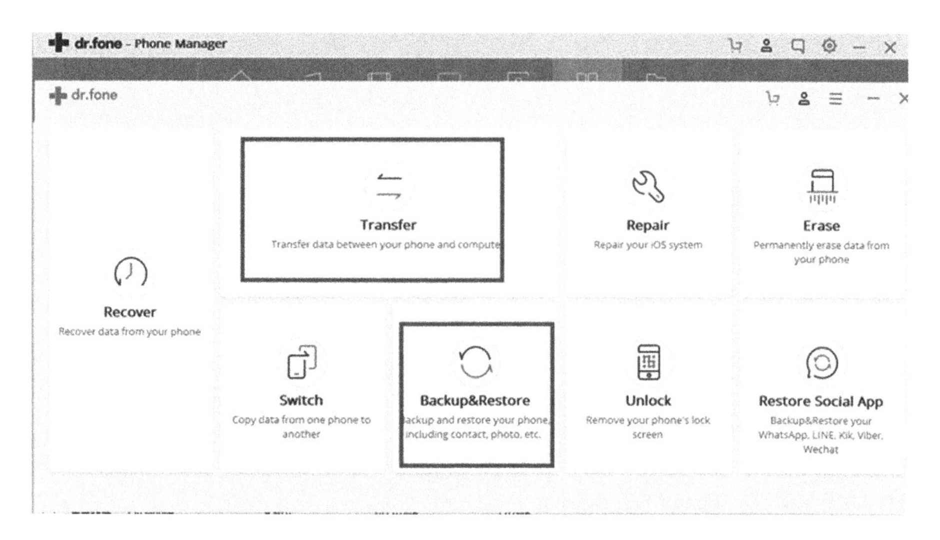

FIGURE 5.10 Dr. Fone transfer data.

Most important will be the ability to copy data from the phone to your forensics workstation. As you can see, Dr. Fone supports this ability. There is the backup option and the transfer option shown in Figure 5.10.

OXYGEN FORENSICS

Oxygen Forensics is known primarily for its easy-to-use interface. The initial connection with a mobile device is accomplished via a wizard, thus making it quite easy to use. It does not have all of the features one finds in tools such as Cellebrite, but certainly many more than tools like Dr. Fone. The pricing is in the neighborhood of $7000 per license. The company website is https://www.oxygen-forensic.com/en/. They formally offered two tools: the Detective and the Analyst version. Now they only offer the Detective version. The wizard is very easy to use, and the first step of that wizard is shown in Figure 5.11.

As was stated earlier, the main benefit of Oxygen is a rather easy to use interface. Once the extraction is done, the results are very easy to work with. Figure 5.12 is a screenshot of Oxygen Forensics results for an older phone I use for forensics labs.

As can be seen, there is a substantial amount of data presented in an easy to find interface. It is quite easy to navigate to events, phone books, messages, and many other pieces of data the forensic examiner may have an interest in. One of the more useful features is placing timeline events on a map for easy geolocation. This is shown in Figure 5.13.

In general, Oxygen is a robust tool with a number of interesting features. It is a reasonable option for the professional forensics lab to include. Given the cost of forensics tools, it is recommended that you seek out recommendations from colleagues, and not rely totally on the marketing information from vendors.

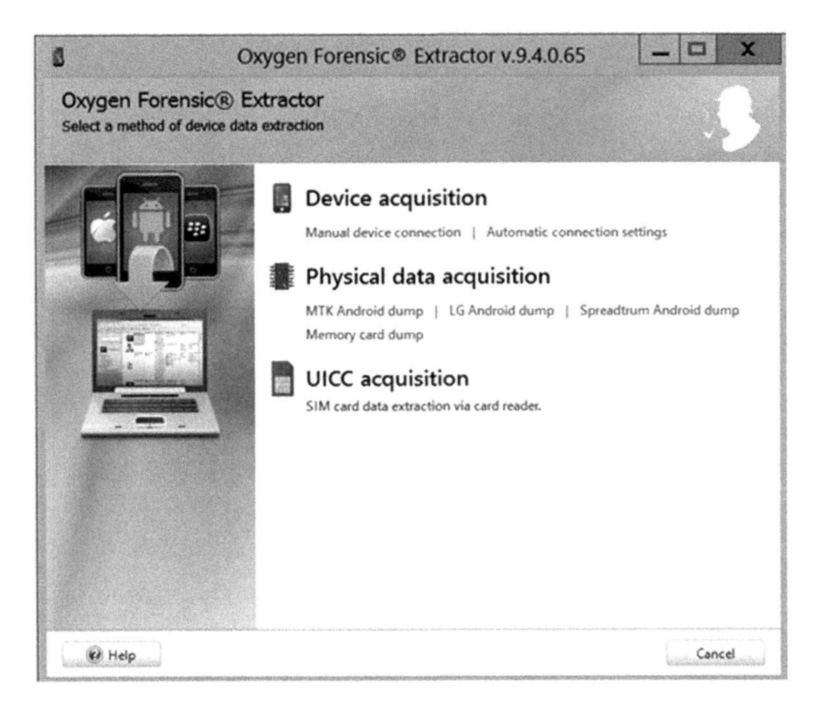

FIGURE 5.11 Oxygen Forensics wizard.

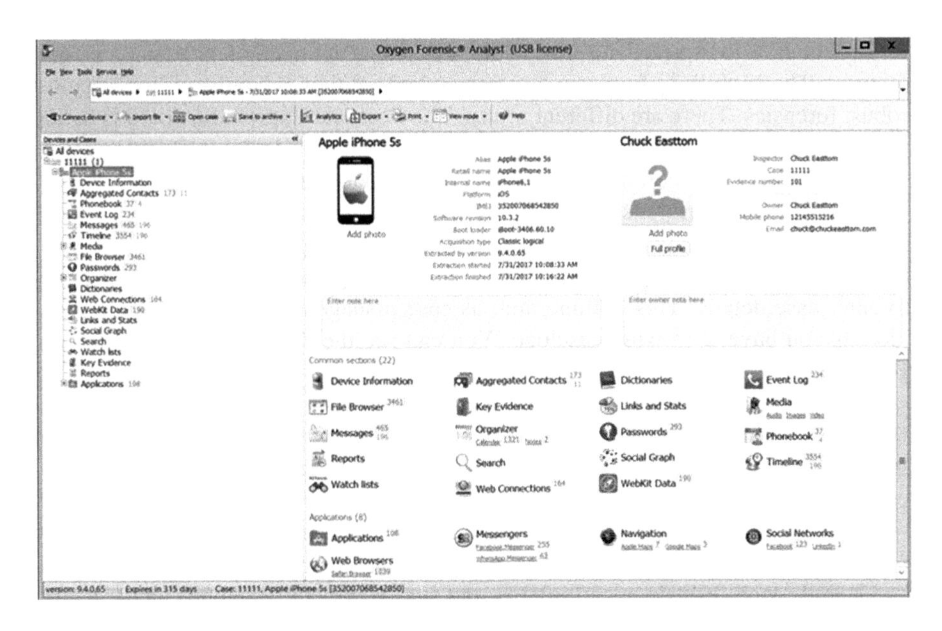

FIGURE 5.12 Oxygen Forensics results.

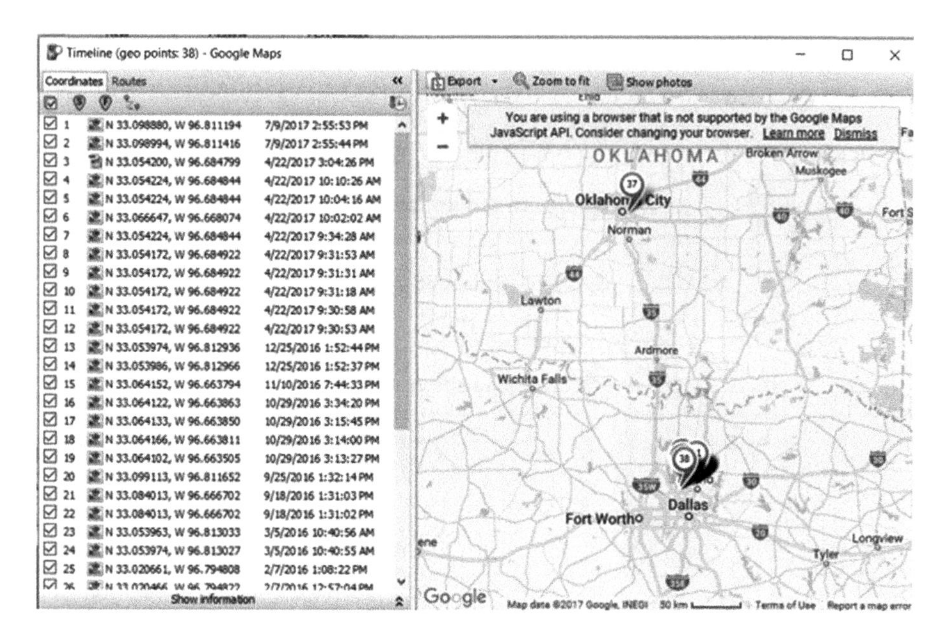

FIGURE 5.13 Oxygen Forensics geolocation.

MobilEdit

MobilEdit is a low-priced forensics tool that has a number of professional-level features. The MobilEdit Forensic Express is the tool they recommend for the most robust forensics. There are different prices. The Forensic Express Standard is $1500 per license, and you need to contact their sales department to get more information on the pricing of Forensic Express Pro. The company website is https://www.mobiledit.com/forensic-express. The starting screen for MobilEdit is shown in Figure 5.14.

Once the device is recognized by the software; the examiner has an opportunity to enter case details. This is important, as case management becomes rather complex as you have a growing caseload. You can see the case information screen in Figure 5.15.

Perhaps the most user-friendly aspect of MobilEdit is the various reporting formats. As you can see in Figure 5.16, the examiner can select multiple formats for the report. The HTML report is easy to navigate. However, the PDF report is often easiest to submit to some third party such as an attorney or other party.

The HTML report is shown in Figure 5.17. This format allows one to simply click on the link on the left in order to navigate to a portion of the report.

Overall, MobilEdit is an affordable and reasonably fully featured tool. It also works with both Apple and Android phones, making it useful on most mobile devices.

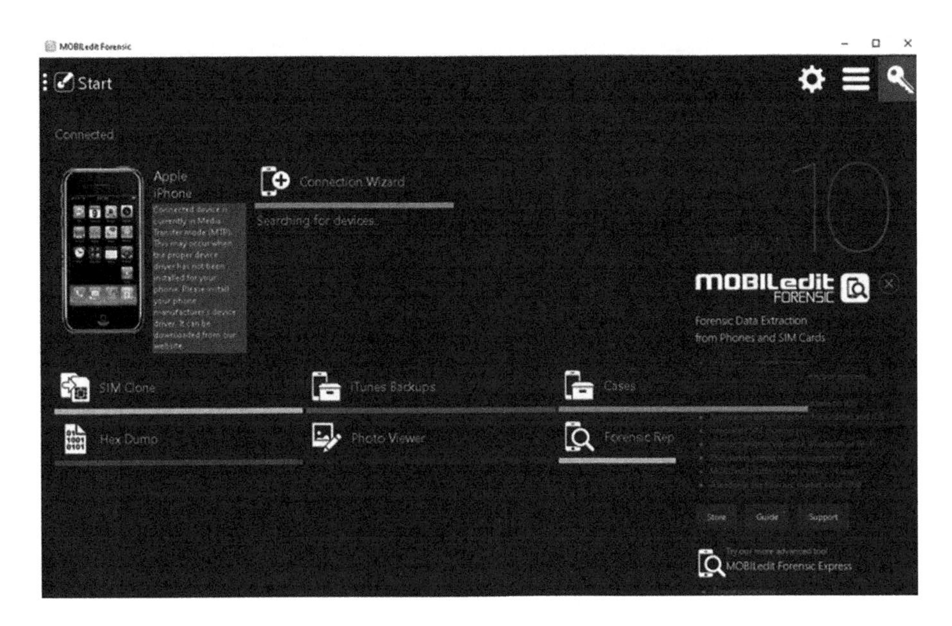

FIGURE 5.14 MobilEdit main screen.

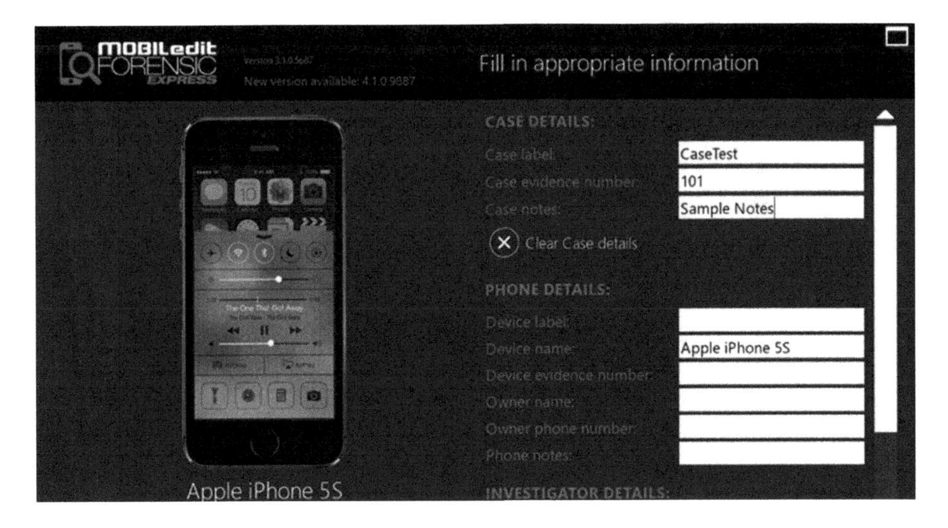

FIGURE 5.15 MobilEdit case details.

Axiom

Axiom is an interesting software package. It can be used to forensically examine mobile devices and PCs. That makes it a very versatile tool. This product is from Magnet Forensics: https://www.magnetforensics.com. They also now have a

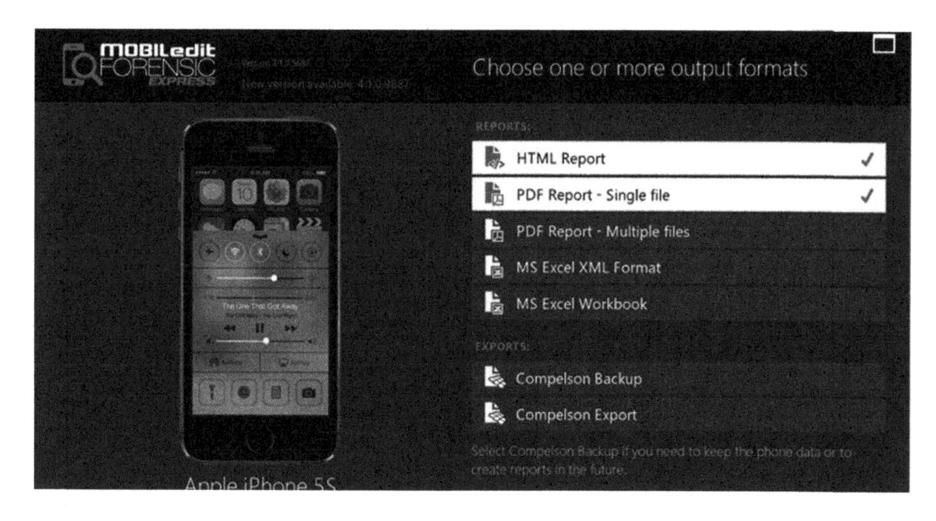

FIGURE 5.16 MobilEdit report options.

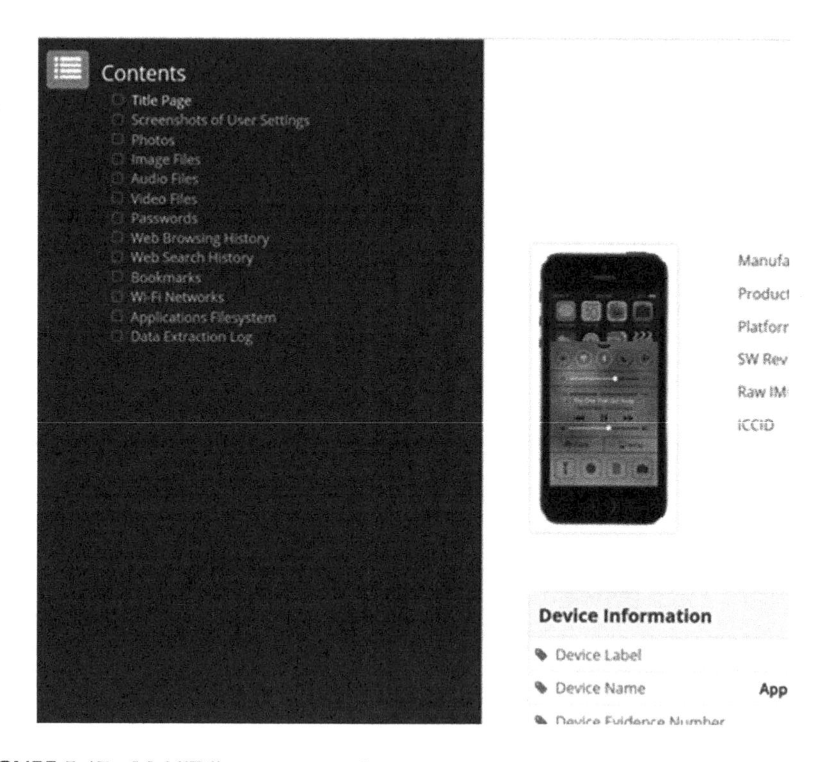

FIGURE 5.17 MobilEdit report example.

cloud-based solution. The precise cost is difficult to ascertain without speaking with a salesperson. However, estimates are in the neighborhood of $7000. The basic case information screen is shown in Figure 5.18.

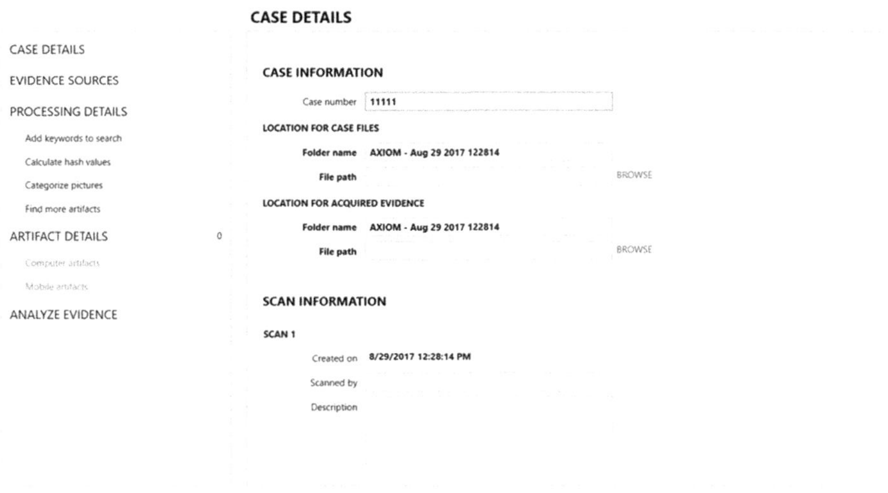

FIGURE 5.18 Axiom case information.

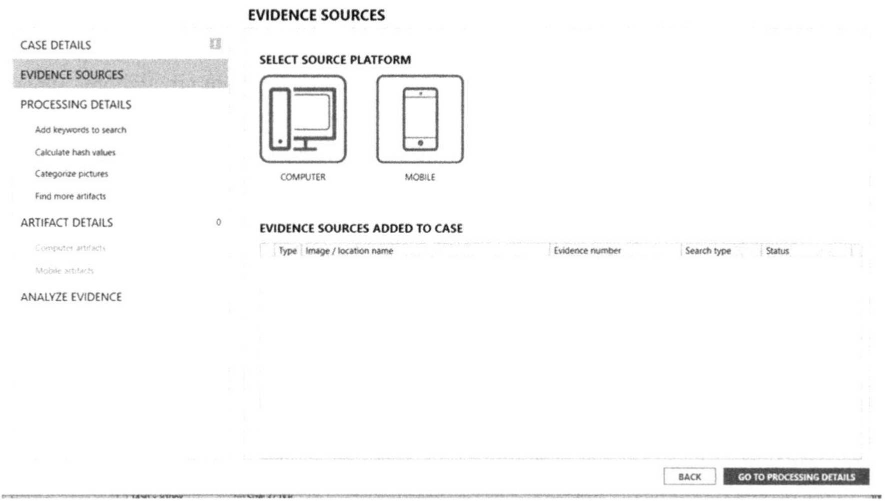

FIGURE 5.19 Axiom selecting case type.

The previously mentioned ability to examine computers and mobile devices is highlighted in Figure 5.19. It is very easy to choose one or the other examiner. This is the most powerful aspect of Axiom forensics.

Axiom does a reasonably good job of extracting data from mobile devices. Figure 5.20 shows a wide array of data that can be retrieved from a mobile device. You can see in the figure, that a lot of information from a wide range of apps.

Overall, Axiom is a well-respected forensics tool that works with Android and iPhone. It is a robust set of features and is relatively easy to use.

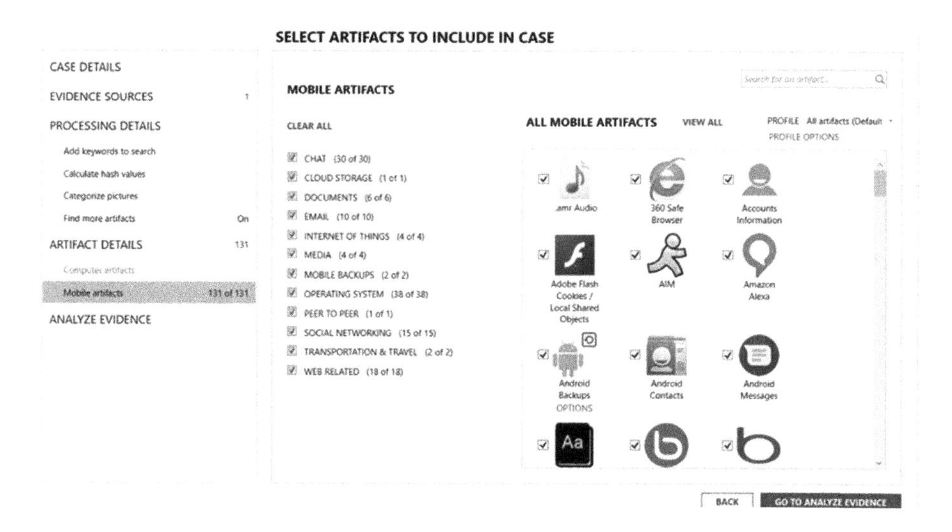

FIGURE 5.20 Axiom data retrieved.

Reincubate

Reincubate makes a number of apps for iOS devices https://reincubate.com/. These are all low priced. while these are not intended for digital forensics, they can accomplish some forensics tasks. For example, extracting data from an iPhone backup. Figure 5.21 shows the data retrieved from a backup of an iPhone.

UltData

This is another tool for analyzing iTunes backups. It may seem like there is a bit too much emphasis on backups in this chapter. However, iPhone security is such that it is not uncommon for an examiner to be unable to circumvent the security and analyze the phone. This will sometimes leave you with the only option being to analyze backup files. UltData allows you to recover from a local iTunes backup, the iCloud, or from a live device. This is shown in Figure 5.22.

The question with any of these tools is simply how much data one gets back. Clearly UltData has an easy-to-use intuitive interface. It is also low priced. Figure 5.23 shows a sample of data retrieved. It can be readily seen that messages, call history, contacts, Safari history, and voicemail are retrieved. But information from WhatsApp, WeChat, and Line are also retrieved.

One interesting feature is the ability to recover old voicemails. For Figure 5.24 a phone was used that has been recently purchased, with a SIM card from an older phone transferred. Furthermore, this is a phone on which old voicemails are periodically deleted. However, one can readily see a voicemail from 3 years past still recovered.

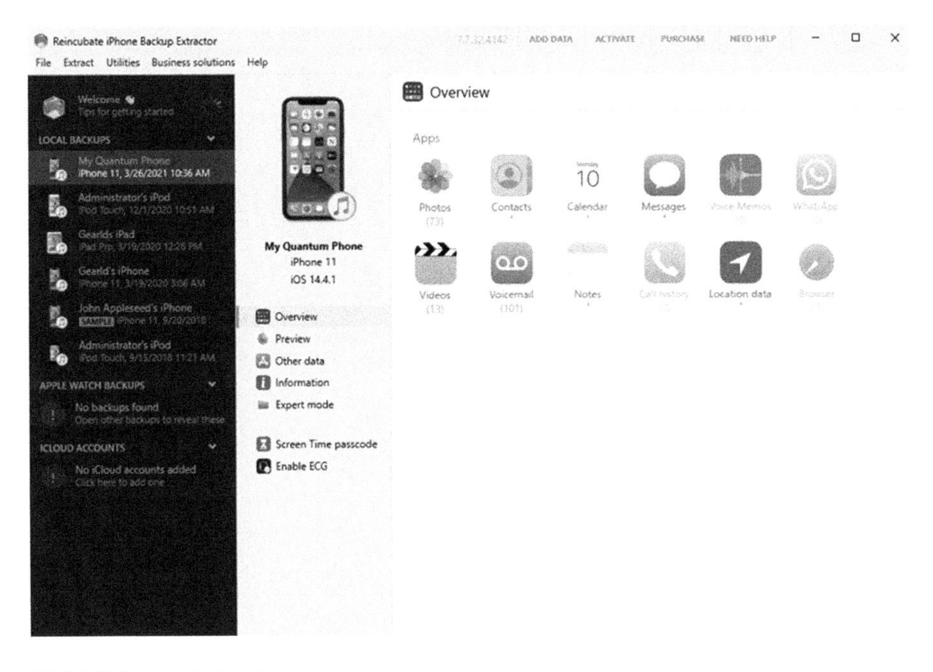

FIGURE 5.21 Reincubate backup extractor.

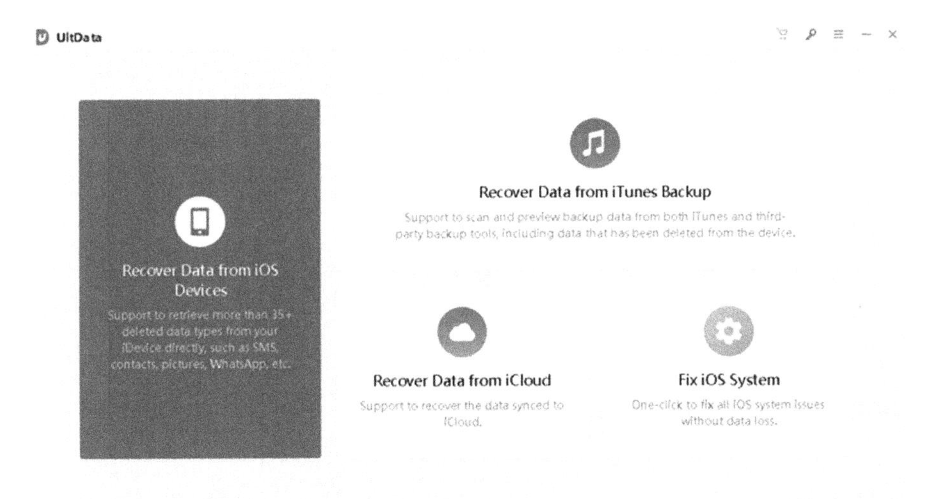

FIGURE 5.22 UltData.

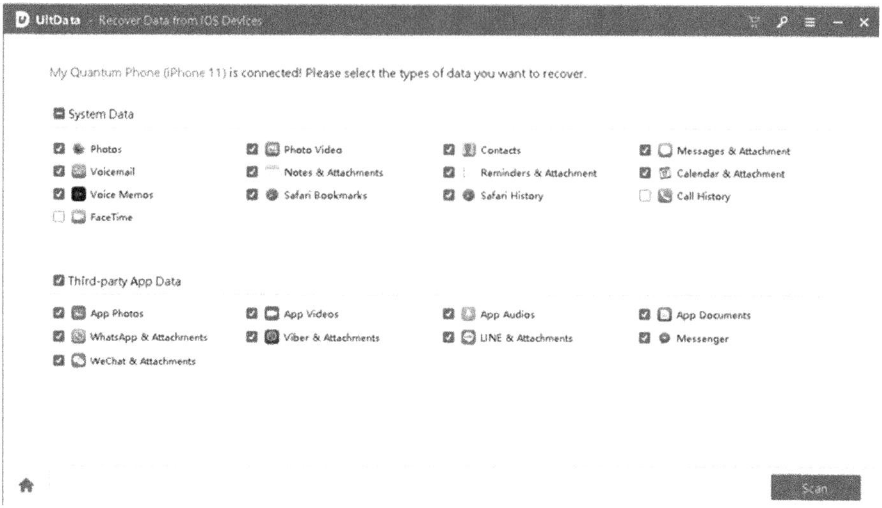

FIGURE 5.23 UltData data retrieved.

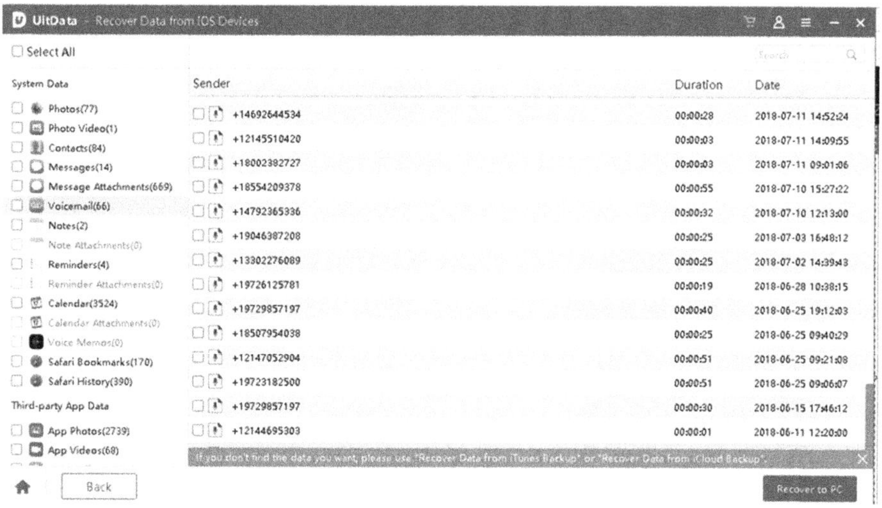

FIGURE 5.24 UltData voicemail recovery.

iCLOUD

If you are unable to access the iPhone, it is possible to get much of the data from the suspect's iCloud account. As of 2018 850 million customers back up to the cloud. Information stored in the iCloud can be retrieved by anyone without having access to a physical device, provided that the original Apple ID and password are known. Often time the username and password can be extracted from the user's computer.

CHAPTER SUMMARY

This chapter has focused on iOS forensics. Chapter 4 should have provided you with a thorough background on iOS operating system and hardware. However, this chapter demonstrated how to extract evidence from the actual device. Specific iOS techniques and concerns were explained. A wide range of tools, with a broad assortment of prices and capabilities, were introduced. This should provide the reader a good general understanding of the tools available.

CHAPTER 5 ASSESSMENT

1. How many communication apps can most forensics tools acquire data from?
   a. Less than six
   b. All apps
   c. A few
   d. Most
2. According to SWGDE, what is the third level of forensic examination?
   a. Manual
   b. Logical
   c. File System
   d. Physical
3. A method that requires disassembling the phone and accessing the circuit board.
   a. JTAG
   b. Chip-Off
   c. MicroRead
   d. Physical
4. What is the best way to describe what must be in an expert report?
   a. The tools used.
   b. The techniques used.
   c. The process used.
   d. The basis and reasons for all opinion
5. The _____ is a unique address that identifies the access point/router that creates the wireless network.
   a. MAC Address
   b. SSID
   c. BSSID
   d. WAP Name

CHAPTER 5 LABS

For these labs you will need an Apple phone. Even an older model, or one with defects such as a broken screen will be adequate. As long as it powers on, that is the issue. Such phones can be found used on eBay, Amazon.com, and phone repair stores for approximately $20 US Dollars. If you have an iPhone you can use that. These labs will not hurt your device.

Lab 5.1

Use either https://imazing.com ($49.99) or https://www.imyfone.com (49.99). If in a classroom setting students may partner to lower the cost even more.

Extract all the data you can from your iPhone device.

You should write a lab report as if this were an actual case. Describe your processes in detail, what you found, any issues you encountered. Consider the guidelines for forensics reports discussed in this chapter.

NOTES

1 https://www.ncjrs.gov/pdffiles1/nij/199408.pdf
2 https://www.sans.org/blog/intro-to-report-writing-for-digital-forensics/
3 National Research Council. (2011). *Reference manual on scientific evidence*. National Academies Press.
4 https://www.cellebrite.com/en/product/
5 Troutman, C., & Mancha, V. (2020). Mobile forensics. In *Digital Forensic Education* (pp. 175–201). Cham: Springer.
6 Gangula, M. R. (2019). Overcoming Forensic Implications with Enhancing Security in iOS.
7 Epifani, M., & Stirparo, P. (2016). *Learning iOS forensics*. Packt Publishing Ltd.

6 Android Forensics

Given the ubiquitous nature of Android devices, Android forensics is essential. Forensics for Android includes approaches, techniques, and tools. Many tools are covered in this chapter, but also extensive use of the Android debugging bridge (ADB). The focus is on free tools or low cost tools.

In this chapter, some material from chapter five is repeated. This is due to the fact that some readers will only be interested in iOS forensics, or in Android forensics, and some of the information overlaps. One reason why mobile devices provide so much evidence is the ubiquitous nature of such devices. Many people would not think of going anywhere without their smartphone. Others carry a tablet with them everywhere. More and more of our lives are conducted on our devices. From ordering food to communicating with friends. Because of the pervasive nature of mobile devices, mobile forensics is important in all types of investigations.

Items you should attempt to recover from a mobile device include the following:

- Call history
- Emails, texts, and/or other messages
- Photos and videos
- Phone information
- Global positioning system (GPS) information
- Network information

The call history lets you know who the user has spoken to and for how long. This information is often important to all types of investigations. Of course, this information is easily erasable, but many users don't erase their call history. Or perhaps the suspect intended to delete this data and simply did not get to it yet. Usually call history does not provide direct evidence of a specific crime, so much as it provides supporting evidence and general intelligence about the suspect's activity.

However, in a cyberstalking case, the call history can be of central importance. The call history can show a pattern of contact with the victim. However, in other cases, it provides only circumstantial evidence. For example, if John Smith is suspected of drug dealing and his call history shows a pattern of regular calls to a known drug supplier, by itself this is not adequate evidence of any crime. However, it aids the investigators in getting an accurate picture of the entire situation.

Smartphones also allow text messages and email. There is also a wide range of chat apps such as Snapchat, Viber, WhatsApp, WeChat, Signal, etc. Many phone forensics tools will retrieve information from some apps. However, given there are hundreds of apps, no tool could possibly retrieve data from all of them. In fact, most tools only get data from a few dozen. In chapter 8 we will discuss SQLite forensics, which will allow you to extract data from the app databases directly.

DOI: 10.1201/9781003118718-6

Photos and videos can provide direct evidence of a crime. In the case of child pornography cases, the relevance is obvious. But there are certainly other crimes where the evidence is found in photos and videos on the suspect's own device. The breach of the U.S. Capitol on January 6, 2021 is a case in point. Many of these who breached the Capitol building took selfies. In many cases uploading those to social media platforms.

This is in no way anomalous. It may surprise you to find that it is quite common for at least some criminals to have incriminating photos and videos on their devices. There are cases of gang members with photos depicting them with illegal firearms and narcotics, thieves videoing themselves during a theft, and many more similar examples. It is always worthwhile to examine photos and videos on a phone.

Information about the phone should be one of the first things you document in your investigation. This will include model number, IMEI number, serial number of the SIM card, operating system, and other similar information. The more detailed, descriptive information you can document, the better. It is important to fully document the details of the phone.

Global positioning system information has become increasingly important in a variety of cases. GPS information, even if it is not exact. can determine if a suspect was in a particular area at the time of the crime. That data can also provide an alibi. If a crime is committed, and phone GPS indicates the person suspected of the crime was actually 100 miles away at the time, this is a compelling alibi. GPS information has begun to play a significant role in contentious divorces, for instance. If someone suspects a spouse of being unfaithful, determining that the spouse's phone and his or her car were at a specific motel when he or she claimed to be at work can be important.

It should be noted that with older phones, the phone did not have true GPS. Rather than use GPS satellites to determine location (which is the most reliable method), they instead would use triangulation of signal strength with various cell towers. This could lead to inaccuracies of up to 50 to 100 feet. However, this situation has changed. Most modern phones and/or the apps on the phone use true GPS for much more accurate data.

The use of Wi-Fi along with GPS will improve accuracy of GPS. The reason for this is that various organizations, including Google, track the Basic Service Set Identifier (BSSID) used by wireless routers, and correlate it with physical addresses. The BSSID is a unique address that identifies the access point/router that creates the wireless network. To identify access points and their clients, the access points MAC address is used. This implies that if your phone connects to a wireless access point, then even with no other data, the phone's location can be pinpointed to within a reasonably close distance of that access point. Network information is also important. Many phones will store the various Wi-Fi hotspots they have connected to. This data could give you an indication of where the phone has been. For example, if the phone belongs to someone suspected of stalking a victim, and the suspect's phone network records show he or she has frequently been using Wi-Fi networks in close proximity to the victim's home, this can be important evidence.

Regardless of the make or model of a mobile phone, these mobile devices are a repository of invaluable data. As was stated earlier in this chapter, so much of our lives is tied to our mobile devices.

FORENSIC PROCEDURES

There are various standards that provide guidelines for forensic examinations. Mobile forensics is no exception. One standard procedure that you should absolutely follow is to put the phone in airplane mode while working with it. You do not want anyone to be able to remotely access the device. It is also important that you make as few changes on any device you are examining, and airplane mode will assist you with that.

One group that is at the forefront of digital forensics procedures is the Scientific Working Group on Digital Evidence (https://www.swgde.org/). SWGDE provides guidance on many digital forensics topics. Related to mobile device forensics, SWGDE provides a general overview of the types of phone forensic investigations:

> Mobile Forensics Pyramid – The level of extraction and analysis required depends on the request and the specifics of the investigation. Higher levels require a more comprehensive examination, additional skills and may not be applicable or possible for every phone or situation. Each level of the Mobile Forensics Pyramid has its own corresponding skill set. The levels are:
>
> 1. Manual – A process that involves the manual operation of the keypad and handset. display to document data present in the phone's internal memory.
> 2. Logical – A process that extracts a portion of the file system.
> 3. File System – A process that provides access to the file system.
> 4. Physical (Non-Invasive) – A process that provides physical acquisition of a phone's data without requiring opening the case of the phone.
> 5. Physical (Invasive) – A process that provides physical acquisition of a phone's data requiring disassembly of the phone providing access to the circuit board (e.g., JTAG).
> 6. Chip-Off – A process that involves the removal and reading of a memory chip to conduct analysis.
> 7. MicroRead – A process that involves the use of a high-power microscope to provide a physical view of memory cells.

Regardless of the tool you choose, you should ensure that the examination chosen is adequate for your forensic needs. Levels 5 and 6, JTAG and MicroRead, are not as common as the other methods. While level 1, manual, is usually not detailed enough. Level 4 is the most common used, but sometimes a level 2 or 3 can be sufficient.

It is also important to have the appropriate tools for the examination. The United States National Institute of Standards (NIST) provides guidance on this issue. The NIST-sponsored CFTT – Computer Forensics Tool Testing Program (http://www.cftt.nist.gov/) provides a measure of assurance that the tools used in the investigations of computer-related crimes produce valid results. Testing includes a set of core requirements as well as optional requirements. It is a good idea to refer to these standards when selecting a tool.

NIST also provides general guidelines on how to write a report for a mobile device forensic report. The guidelines are of what to include are:

- Descriptive list of items submitted for examination, including serial number, make, and model.
- Identity and signature of the examiner.
- The equipment and setup used in the examination.
- Brief description of steps taken during examination, such as string searches, graphics image searches, and recovering erased files.
- Supporting materials, such as printouts of particular items of evidence, digital copies of evidence, and chain of custody documentation.
- Details of findings:
  - Specific files related to the request.
  - Other files, including deleted files, that support the findings.
  - String searches, keyword searches, and text string searches.
  - Internet-related evidence, such as website traffic analysis, chat logs, cache files, email, and news group activity.
  - Graphic image analysis.
  - Indicators of ownership, which could include program registration data.
  - Data analysis.
  - Description of relevant programs on the examined items.
  - Techniques used to hide or mask data, such as encryption, steganography, hidden attributes, hidden partitions, and file name anomalies.
- Report conclusions

The United States Department of Justice[1] states

> The examiner is responsible for completely and accurately reporting his or her findings and the results of the analysis of the digital evidence examination. Documentation is an ongoing process throughout the examination. It is important to accurately record the steps taken during the digital evidence examination.

The well-respected SANS institute publishes guidelines[2] on digital forensics reports. They state a digital forensics report should include overview/case summary, details on the forensic acquisition and exam, details of all steps taken, etc. The Reference Manual on Scientific Evidence: Third Edition[3] states "Under Federal Rule of Civil Procedure 26(a)(2)(B)(i), the expert report must contain the basis and reasons for all opinions expressed, and certainly the expectation is that oral testimony will do the same." An inadequate report can lead to the report and the associated testimony being excluded at trial.

The important thing to remember when creating any forensics report is that it is quite difficult to put too much detail in. It is much better to have information that is ultimately not necessary, than to wish you had captured data later. When forensics cases go to court, there may be an opposing forensic expert. The idea of a forensic report is that it is so detailed, that any competent forensic examiner can recreate the steps you took and either verify or refute your conclusions. This means that your report should quite literally create a roadmap of your investigation, and one that any competent examiner can follow.

ADB

Android Debugging Bridge was introduced in chapter 4. In this section it will be used specifically to conduct forensic examinations of Android devices. If you require a refresher on ADB, please revisit chapter 4.

When you start an adb client, the client first checks whether there is an adb server process already running. If there isn't, it starts the server process. When the server starts, it binds to local TCP port 5037 and listens for commands sent from adb clients – all adb clients use port 5037 to communicate with the adb server. When a device is connected to a computer that has ADB, the first step is to list all connected devices.

One of the first tasks you should do is to make a backup of the device. The general format of the command is: adb backup -all -f backup. A specific example would be:

adb backup -all -f c:\phonebackup\. You can see a backup in Figure 6.1.

This will create a backup of all the data on the phone that is user accessible. In some cases, you may even want to restore that backup to a lab phone in order to conduct your investigation. That is not widely practiced in forensics but would provide a guarantee that you do not alter anything on the actual device in question. However, even without working on a lab phone, you do have the backup which can be used to confirm you did not accidentally alter any data or settings on the suspect phone.

Once one has made a backup, the next step will be to actually begin the forensic examination on either a lab phone with the backup restored, or the suspect phone. Either way, one of the first commands is to list contents in the Linux shell:

adb shell

ls

This provides a view of what is actually on the device. You will want to navigate to directories and explore their contents. You will find some directories have permission denied. Those are system directories that can only be accessed on a rooted

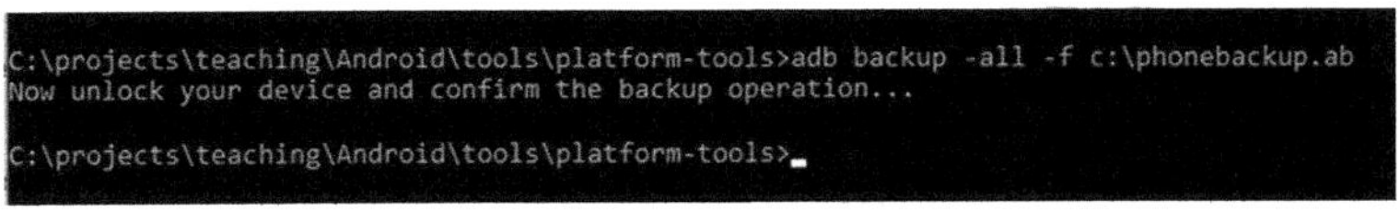

```
C:\projects\teaching\Android\tools\platform-tools>adb backup -all -f c:\phonebackup.ab
Now unlock your device and confirm the backup operation...

C:\projects\teaching\Android\tools\platform-tools>_
```

FIGURE 6.1 ADB backup.

```
nobleltespr:/data $ ls
ls: .: Permission denied
1|nobleltespr:/data $ su ls
nobleltespr:/data # ls
DownFilters    biometrics    hs20        nfc
adb            bootchart     iq_archive  ota
anr            camera        knox        otp
app            clipboard     local       overlays
app-asec       custom_image  log         privatemode
app-ephemeral  dalvik-cache  lost+found  property
app-lib        data          media       resource-cache
app-private    drm           mediadrm    security
app_fonts      enc_user      misc        snd
backup         firmware      misc_ce     ss
bcmnfc         fota          misc_de     ss_conn_daemon.pid
```

FIGURE 6.2 ADB is command.

phone. Figure 6.2 shows the ls content in the data directory. Sometimes, you can try su ls (or su any command) and get the command to work.

There are specific directories you will want to look for evidence in. These are described in the following subsections.

/DATA

This partition contains the user's data like your contacts, sms, settings, and all android applications that you have installed. Should you perform factory reset on your device, this partition will be wiped out. This directory is likely to have much of the evidence you are seeking. Some of it will be in SQLite Databases which we will explore in chapter 8.

The sub=directory /data/app contains apps installed but not by the vendor. /data/data/<package>/databases has the databases for specific apps. This is where you will find the SQLite database we will examine in chapter 8.

/CACHE

This is the partition where Android stores frequently accessed data and app components.

Wiping the cache doesn't affect your personal data but simply gets rid of the existing data there, which gets automatically rebuilt as you continue using the device. This can sometimes reveal interesting evidence in a case.

/MISC

This partition contains miscellaneous system settings in form of on/off switches. These settings may include CID (Carrier or Region ID), USB configuration and certain hardware settings etc.

/MNT

If there is an SD card, internal or external, you will find its stat here. This is a very important place to check as it could potentially have quite a bit of evidence. While there can be variations between models, in general the subdirectories you will find are listed here:

/mmt/asec (encrypted apps)
/mmt/emmc (internal SD Card)
/mmt/sdcard (external/Internal SD Card)
/mmt/sdcard/external_sd (external SD Card)

Commands to Execute

It is also important to get information regarding the phone itself, not just its content. There are a series of commands that can be executed for this purpose:

```
adb shell getprop ro.product.model
adb shell getprop ro.build.version.release
adb shell getprop ro.serialno
adb shell getprop ro.board.platform
adb shell getprop ro.build.description
adb shell getprop ro.product.locale.language
adb shell getprop ro.product.locale.region
adb shell getprop ro.board.platform
adb shell getprop ro.build.version.codename
```

Running the preceding shell commands and documenting the output, is an excellent way to document the system you are analyzing. It provides the starting point for your investigation. Then you will begin to navigate through the various folders utilizing a combination of the ls and cd commands. The ls will list contents, while cd will change directories. There are also variations of the ls command you will find useful. The first such variation is *ls -l* which shows file or directory, size, modified date and time, file or folder name and owner of file and its permission. This is shown in Figure 6.3

All commands in Linux have a number of flags, like the -l flag shown in Figure 6.3. These alter the behavior of the command in question, producing different outputs. Here are a few important ls flags you will want to keep in mind:

Show hidden files: *ls -a*
Show sizes in a human readable format: *ls -lh*
Recursively show sub directories: *ls -R*
Show last modification date: *ls -ltr*
See UID or GID of files: *ls -n*
Sort by time and date: *ls -t*

```
hero2qlteue:/ $ ls -l
total 9688
dr-xr-xr-x 97 root   root          0 2016-06-08 21:44 acct
drwxrwx--x  2 system system       40 2016-06-08 21:44 bt_firmware
lrwxrwxrwx  1 root   root         50 1969-12-31 19:00 bugreports -> /data/user_de/0/com.andro
drwxrwx---  6 system cache      4096 2016-04-11 17:37 cache
drwxrwx--x  3 system system     4096 2015-12-31 19:04 carrier
drwxr-xr-x  2 root   root          0 2016-06-08 21:44 config
lrwxrwxrwx  1 root   root         17 1969-12-31 19:00 d -> /sys/kernel/debug
drwxrwx--x 61 system system     4096 2016-06-08 21:44 data
-rw-r--r--  1 root   root       1345 1969-12-31 19:00 default.prop
drwxr-xr-x 17 root   root       4760 2016-02-27 23:07 dev
drwxr-xr-x  3 root   root       4096 1969-12-31 19:00 dsp
drwxrwx--x 26 radio  system     4096 2015-12-31 19:06 efs
lrwxrwxrwx  1 root   root         11 1969-12-31 19:00 etc -> /system/etc
-rw-r--r--  1 root   root     398770 1969-12-31 19:00 file_contexts.bin
dr-xr-x---  3 system system    16384 1969-12-31 19:00 firmware
dr-xr-x---  4 system system    16384 1969-12-31 19:00 firmware-modem
-rw-r-----  1 root   root       1351 1969-12-31 19:00 fstab.qcom
-rwxr-x---  1 root   root    3447376 1969-12-31 19:00 init
-rwxr-x---  1 root   root       4577 1969-12-31 19:00 init.carrier.rc
-rwxr-x---  1 root   root       3301 1969-12-31 19:00 init.class_main.sh
-rwxr-x---  1 root   root       3763 1969-12-31 19:00 init.container.rc
-rwxr-x---  1 root   root       1693 1969-12-31 19:00 init.environ.rc
-rwxr-x---  1 root   root       1730 1969-12-31 19:00 init.mdm.sh
-rwxr-x---  1 root   root      28931 1969-12-31 19:00 init.msm.usb.configfs.rc
-rwxr-x---  1 root   root       7054 1969-12-31 19:00 init.qcom.class_core.sh
-rwxr-x---  1 root   root      11561 1969-12-31 19:00 init.qcom.early_boot.sh
-rwxr-x---  1 root   root       3468 1969-12-31 19:00 init.qcom.factory.rc
-rwxr-x---  1 root   root      35141 1969-12-31 19:00 init.qcom.rc
-rwxr-x---  1 root   root       2056 1969-12-31 19:00 init.qcom.sensors.sh
-rwxr-x---  1 root   root      12089 1969-12-31 19:00 init.qcom.sh
-rwxr-x---  1 root   root       2962 1969-12-31 19:00 init.qcom.syspart_fixup.sh
-rwxr-x---  1 root   root     102004 1969-12-31 19:00 init.qcom.usb.rc
-rwxr-x---  1 root   root       9920 1969-12-31 19:00 init.qcom.usb.sh
-rwxr-x---  1 root   root      69052 1969-12-31 19:00 init.rc
-rwxr-x---  1 root   root        544 1969-12-31 19:00 init.rilcarrier.rc
-rwxr-x---  1 root   root       2068 1969-12-31 19:00 init.rilchip.rc
```

FIGURE 6.3 ADB is -l command.

If you are working with the actual suspect phone, it will be useful to have the running processes. The Linux ps command can provide that. The ps command is shown in Figure 6.4.

As you may already be able to surmise, the ps command also has a number of interesting flags that can alter the input. Some common flags you are likely to find useful are listed here:

Show active processes: ps -A or ps -E
See processes running with root privileges: ps -U root -u root
See the process tree: ps -e –forest
See the process tree for a specific process: ps -f –forest -c sshd
See child processes for a specific process: ps – C sshd
See all running processes: ps –r

These are only a few options, but they are important. It should be quite obvious how these commands might be of interest in a forensic investigation. With any of the Linux shell commands you can always type the command followed by -? or -help to get a list of all the possible flags.

```
[2J[Hhero2qlteue:/ $ ps
USER     PID   PPID  VSIZE  RSS   WCHAN              PC  NAME
root     1     0     31428  1640  SyS_epoll_ 0000000000 S /init
root     2     0     0      0       kthreadd 0000000000 S kthreadd
root     3     2     0      0     smpboot_th 0000000000 S ksoftirqd/0
root     6     2     0      0     worker_thr 0000000000 S kworker/u8:0
root     7     2     0      0     rcu_gp_kth 0000000000 S rcu_preempt
root     8     2     0      0     rcu_gp_kth 0000000000 S rcu_sched
root     9     2     0      0     rcu_gp_kth 0000000000 S rcu_bh
root     10    2     0      0     smpboot_th 0000000000 S migration/0
root     11    2     0      0     smpboot_th 0000000000 S migration/1
root     12    2     0      0     smpboot_th 0000000000 S ksoftirqd/1
root     14    2     0      0     worker_thr 0000000000 S kworker/1:0H
root     15    2     0      0     smpboot_th 0000000000 S migration/2
root     16    2     0      0     smpboot_th 0000000000 S ksoftirqd/2
root     17    2     0      0     worker_thr 0000000000 S kworker/2:0
root     18    2     0      0     worker_thr 0000000000 S kworker/2:0H
root     19    2     0      0     smpboot_th 0000000000 S migration/3
root     20    2     0      0     smpboot_th 0000000000 S ksoftirqd/3
root     23    2     0      0     rescuer_th 0000000000 S khelper
root     24    2     0      0     rescuer_th 0000000000 S netns
root     25    2     0      0     rescuer_th 0000000000 S perf
root     26    2     0      0     rescuer_th 0000000000 S smd_channel_clo
root     27    2     0      0     kthread_wo 0000000000 S dsps_smd_trans_
root     28    2     0      0     kthread_wo 0000000000 S lpass_smd_trans
root     29    2     0      0     kthread_wo 0000000000 S mpss_smd_trans_
root     30    2     0      0     kthread_wo 0000000000 S wcnss_smd_trans
root     31    2     0      0     kthread_wo 0000000000 S rpm_smd_trans_g
root     32    2     0      0     watchdog_k 0000000000 S msm_watchdog
root     33    2     0      0     worker_thr 0000000000 S kworker/1:1
root     34    2     0      0     rescuer_th 0000000000 S rpm_requests
```

FIGURE 6.4 ADB ps command.

```
:\platform-tools>adb shell netstat
ctive Internet connections (w/o servers)
roto Recv-Q Send-Q Local Address          Foreign Address         State
ctive UNIX domain sockets (w/o servers)
roto RefCnt Flags       Type       State         I-Node Path
nix  2      [ ]         STREAM                   14336 /dev/socket/thermal-send-client
nix  2      [ ]         DGRAM                    18471 @suilst
nix  2      [ ]         STREAM                   19461 /dev/socket/thermal-recv-client
nix  2      [ ]         STREAM                   19464 /dev/socket/thermal-recv-passive-client
nix  2      [ ]         DGRAM                    52286 /data/misc/wifi/sockets/wlan0
nix  2      [ ]         DGRAM                    58163 @suisvc
nix  3      [ ]         STREAM     CONNECTED     17260 /dev/socket/qmux_radio/qmux_client_socket   10
nix  3      [ ]         STREAM     CONNECTED     18838 /dev/socket/qmux_radio/qmux_client_socket   9
nix  2      [ ]         DGRAM                    80302 /dev/socket/mtp/mtp_event_socket
nix  4      [ ]         DGRAM                    47022 /dev/socket/wpa_wlan0
nix  2      [ ]         SEQPACKET                19649 /dev/socket/cellgeofence
nix  2      [ ]         DGRAM                    51552 /data/misc/wifi/sockets/wpa_ctrl_1586-2
nix  2      [ ]         DGRAM                    51553 /data/misc/wifi/sockets/wpa_ctrl_1586-3
nix  2      [ ]         DGRAM                    79589 /dev/socket/mtp/mtp_sink_socket
nix  2      [ ]         DGRAM                    79591 /dev/socket/mtp/mtp_source_socket
nix  2      [ ]         DGRAM                    20717 /dev/socket/ipacm_log_file
nix  2      [ ]         DGRAM                    16759 @gplisnr
nix  143    [ ]         DGRAM                    13813 /dev/socket/logdw
nix  3      [ ]         STREAM     CONNECTED     53242 /data/misc/location/mq/location-mq-s
```

FIGURE 6.5 ADB netstat command.

Another important command will be netstat. This is a standard command in Linux and in Windows that shows the network status. Are there any processes on the device that are establishing or attempting to establish an external connection? This command will tell you that and additional relevant details. This can be seen in Figure 6.5.

In chapter 4, the pm list packages command was briefly introduced. This command can be quite interesting forensically. You can view any packages installed on the device. If there is spyware, a hidden app, or other interesting apps, this will help you find them. There are several variations of this command you will likely find quite useful.

pm list packages -f See their associated file.
pm list packages -d Filter to only show disabled packages.
pm list packages -e Filter to only show enabled packages.
pm list packages -i See the installer for the packages.
pm list packages -u Also include uninstalled packages.

The command pm list packages -f is shown in Figure 6.6.

Like many commands, this will often fill up several screens. That is not useful for forensic examination. So, the first step is to exit the shell using exit. Then, with any command, you can export the output to a text file. When executing shell commands from outside the shell, you should precede the command with adb shell. This is shown in Figure 6.7.

Now this does not provide much on the screen, but you have just taken what would have been several screens and output that to a text file. Not only does this allow you to retain a copy of the text file, but you can easily search it. You will find that this is often the best way to conduct forensics on an Android phone using adb.

```
←[2J←[Hstarqlteue:/ $ pm list packages -f
package:/system/app/FilterProvider/FilterProvider.apk=com.samsung.android.provider.filterprovider
package:/system/app/RoseEUKor/RoseEUKor.apk=com.monotype.android.font.rosemary
package:/data/app/com.samsung.oh-Q_uzbNB6KVNmCS7c_sabxw==/base.apk=com.samsung.oh
package:/data/app/SamsungIMEv3.3Tyme_Removable/SamsungIMEv3.3Tyme_Removable.apk=com.sec.android.inputmethod.beta
package:/system/app/AutomationTest_FB/AutomationTest_FB.apk=com.sec.android.app.DataCreate
package:/system/priv-app/LedCoverAppStar/LedCoverAppStar.apk=com.samsung.android.app.ledcoverdream
package:/system/priv-app/CtsShimPrivPrebuilt/CtsShimPrivPrebuilt.apk=com.android.cts.priv.ctsshim
package:/system/app/RootPA/RootPA.apk=com.gd.mobicore.pa
package:/system/priv-app/GalaxyAppsWidget_Phone_Dream/GalaxyAppsWidget_Phone_Dream.apk=com.sec.android.widgetapp.samsung
apps
package:/system/priv-app/SmartSwitchAssistant/SmartSwitchAssistant.apk=com.samsung.android.smartswitchassistant
package:/system/priv-app/NSDSWebApp/NSDSWebApp.apk=com.sec.vsim.ericssonnsds.webapp
package:/system/app/SetupWizardLegalProvider/SetupWizardLegalProvider.apk=com.sec.android.app.setupwizardlegalprovider
package:/data/app/com.google.android.youtube-dqcm2CjykOdbYIkyIYd5_A==/base.apk=com.google.android.youtube
package:/system/priv-app/Finder/Finder.apk=com.samsung.android.app.galaxyfinder
package:/system/framework/VZWAPNLib_VZW.apk=com.vzw.apnlib
package:/system/priv-app/NSFusedLocation_v3.5/NSFusedLocation_v3.5.apk=com.sec.location.nsflp2
package:/system/priv-app/ThemeStore/ThemeStore.apk=com.samsung.android.themestore
package:/system/app/ChromeCustomizations/ChromeCustomizations.apk=com.sec.android.app.chromecustomizations
package:/system/priv-app/AODService_v40/AODService_v40.apk=com.samsung.android.app.aodservice
package:/system/priv-app/CocktailBarService_v3.2/CocktailBarService_v3.2.apk=com.samsung.android.app.cocktailbarservice
package:/vendor/overlay/DisplayCutoutEmulationCorner/DisplayCutoutEmulationCornerOverlay.apk=com.android.internal.displa
y.cutout.emulation.corner
package:/system/priv-app/GoogleExtServices/GoogleExtServices.apk=com.google.android.ext.services
package:/system/priv-app/SuwScriptPlayer/SuwScriptPlayer.apk=com.sec.android.app.suwscriptplayer
package:/vendor/overlay/DisplayCutoutEmulationDouble/DisplayCutoutEmulationDoubleOverlay.apk=com.android.internal.displa
y.cutout.emulation.double
```

FIGURE 6.6 Pm list packages -f.

```
D:\platform-tools>adb shell pm listpackages -f > package.txt

D:\platform-tools>
```

FIGURE 6.7 Export command.

```
starqlteue:/sdcard $ cd DCIM
starqlteue:/sdcard/DCIM $ ls
Camera
starqlteue:/sdcard/DCIM $ cd Camera
starqlteue:/sdcard/DCIM/Camera $ ls
20210329_134750.jpg 20210329_134800.jpg
starqlteue:/sdcard/DCIM/Camera $
```

FIGURE 6.8 sdcard/DCIM/.

```
D:\platform-tools>adb pull /sdcard/DCIM/Camera/20210329_134800.jpg
/sdcard/DCIM/Camera/20210329_134800.jpg: 1 file pulled. 12.3 MB/s (1634113 bytes in 0.127s)

D:\platform-tools>adb pull /sdcard/DCIM/Camera/*.*
adb: error: failed to stat remote object '/sdcard/DCIM/Camera/*.*': No such file or directory

D:\platform-tools>adb pull /sdcard/DCIM/
/sdcard/DCIM/: 2 files pulled. 12.3 MB/s (4818806 bytes in 0.373s)

D:\platform-tools>
```

FIGURE 6.9 Pull data.

When searching an Android phone, a great deal of information will be in the /sdcard/ folder. And images, such as pictures are usually in the /DCIM folder. You can see this in Figure 6.8.

Once you have identified an item you wish to pull you can use the adb pull command to pull just that image or the entire directory to your forensics machine. This is shown in Figure 6.9.

Note, particularly for Windows users, you do not use the *.* to get everything. You simply give the path to a directory. If you don't name a file in the directory, ADB will assume you want the entire directory. This makes it quite convenient to pull data from an Android phone.

ANDROID ONLY TOOLS

As with chapter 5, we will spend substantial time on specific tools you can use for Android forensics. We will begin with low-cost tools that are Android specific. These can be obtained at little or no cost and retrieve a great deal of the information you may need in a forensic examination.

Tool All-In-One

This tool will prompt you for a donation, but you can use it for free. It can be downloaded from https://androidfilehost.com/?a=show&w=files&flid=38683. This is essentially a nice GUI for ADB. It does all the things ADB does, but without requiring you to remember all the ADB commands. If it cannot recognize your phone, you can simply choose "generic device." You will see an initial screen as shown in Figure 6.10.

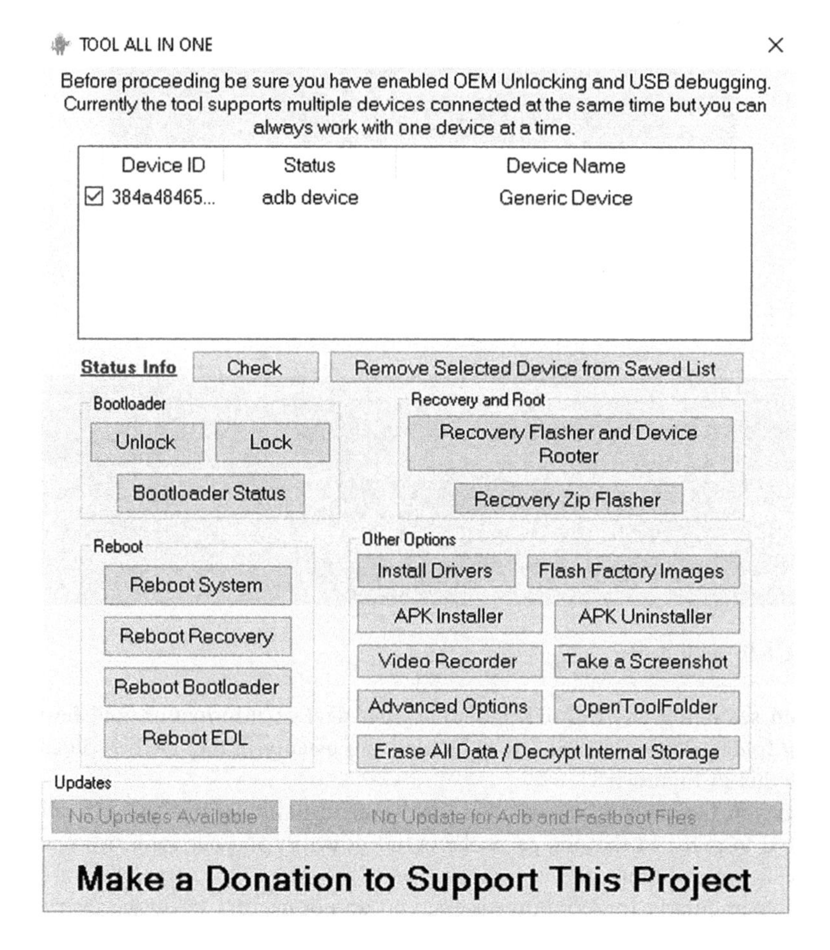

FIGURE 6.10 All in one main screen.

As you can readily see, this is not designed just for forensics. In fact, there are items you probably would never use in a forensic exam such as "Install Drivers," "APK Installer," or "Erase All Data." If you choose the Bootloader unlock, the tool will tell you that this is a fastboot command and ask if you wish to reboot the phone in fastboot mode. You may recall fastboot from chapter 4. The process of unlocking the bootloader is just as difficult and may not work, but this does give you an easy-to-follow GUI to perform the task in. This tool is primarily helpful if you are trying to reboot into some specific mode such as fastboot or EDL.

Android Tools

This is far more versatile than All-In-One and is a free download from https://sourceforge.net/projects/android-tools/. The main screen is shown in Figure 6.11.

FIGURE 6.11 Android tools main screen.

To anyone familiar with ADB this should look quite familiar. You can see various ADB commands simply at the touch of the button. You can also launch a shell console to perform your own Linux commands if you need to. So, you have all the benefits of a GUI, and still can use the Linux shell. There is a tab for fastboot commands that can allow you to attempt to unlock or root the phone. Just as when you perform this manually, there is no guarantee it will work. But you can see the fastboot tab in Figure 6.12.

The advanced tab is very interesting. Among other things it allows you to work with various ADB backup files. This can be quite useful. This tab is shown in Figure 6.13.

We are not describing every feature of this tool for two reasons. The first is that it is automating ADB. And from chapter 4 and the first part of this chapter, you should be comfortable with ADB. Secondly, the interface is so intuitive to use that minimal instruction is needed.

Autopsy

Perhaps the most well-known open-source forensics software is Autopsy. This tool is designed for PC forensics but can analyze mobile phone images. You can download Autopsy for free from https://www.autopsy.com/. Android won't extract

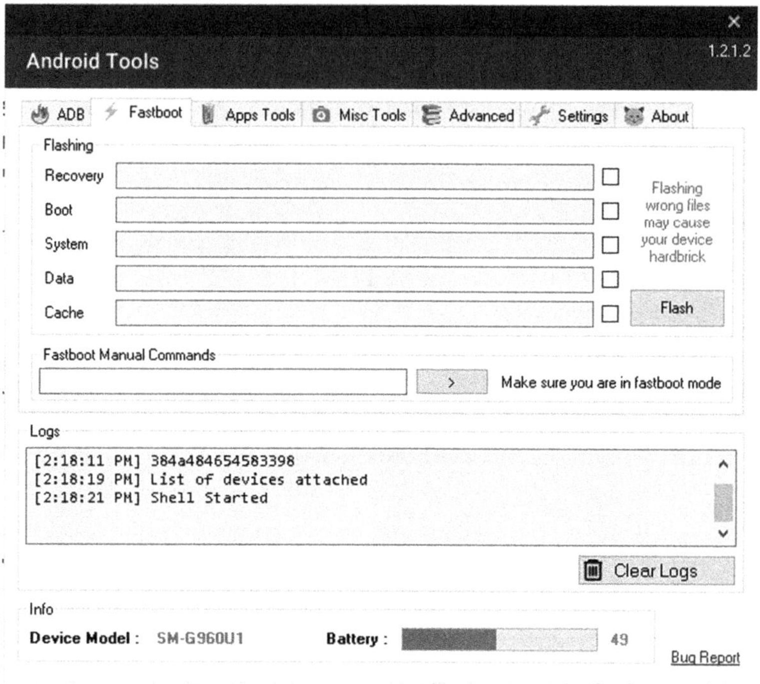

FIGURE 6.12 Android tools fastboot tab.

from your phone, but if you have an image from an Android phone you can examine it with Autopsy. The first step is to add an image. This is shown in Figure 6.14.

You can see in Figure 6.15, that Autopsy can extract quite a bit of information.

Results are shown in Figure 6.16. As you can see the call logs, contacts, messages, GPS track points and more is retrieved.

BitPim

This is an open-source tool you can freely download from https://sourceforge.net/projects/bitpim/. However, it is limited in the phones it can recognize. The basic landing screen is shown in Figure 6.17.

This tool does come with a very useful help file that is easy to navigate and shown in Figure 6.18.

OSAF

Open-Source Android Forensics is a virtual machine you can download for free from https://sourceforge.net/projects/osaftoolkit/. It is Ubuntu Linux with a great many pre-loaded Android forensics tools. The default password is *forensics*. The desktop is shown in Figure 6.19.

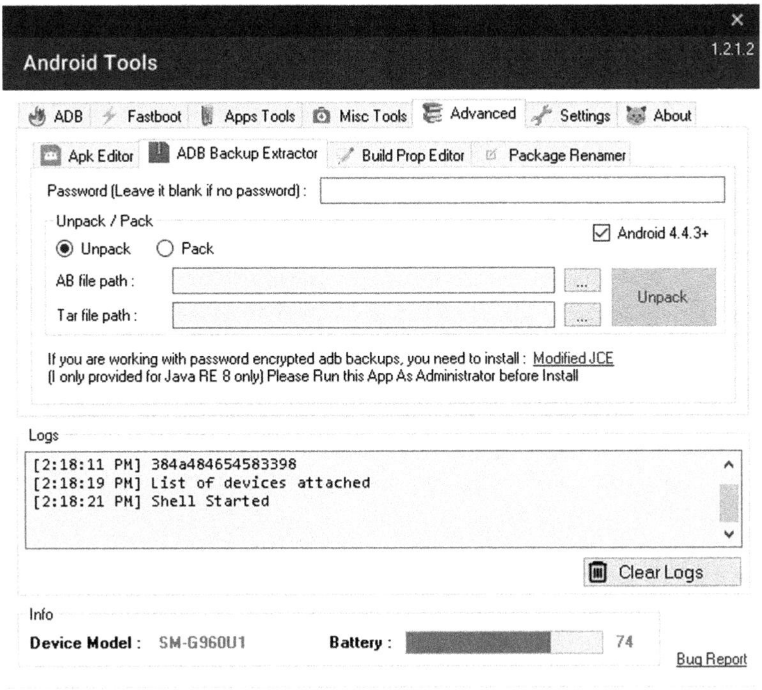

FIGURE 6.13 Android tools advanced tab.

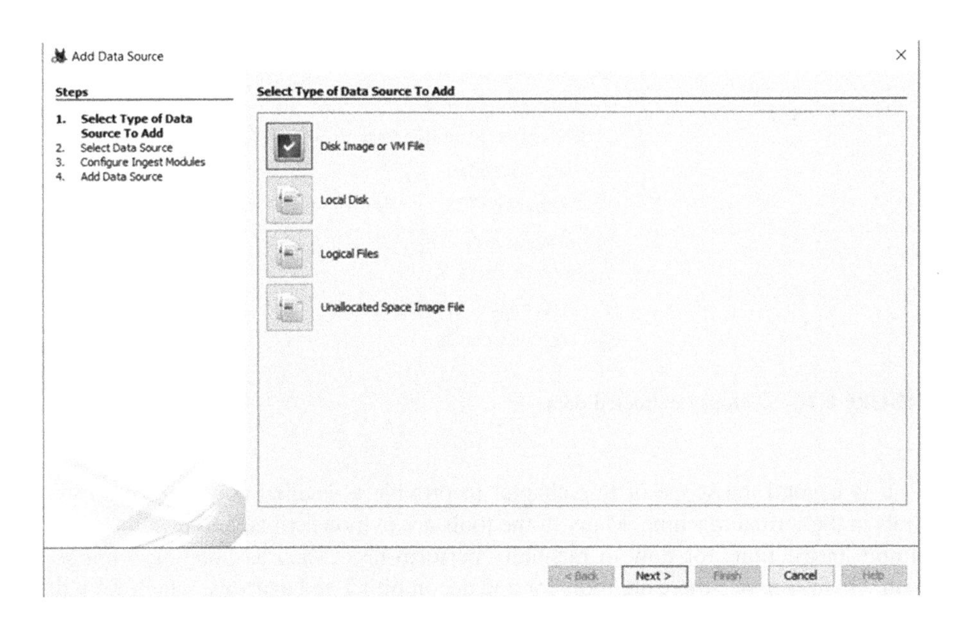

FIGURE 6.14 Autopsy add an image.

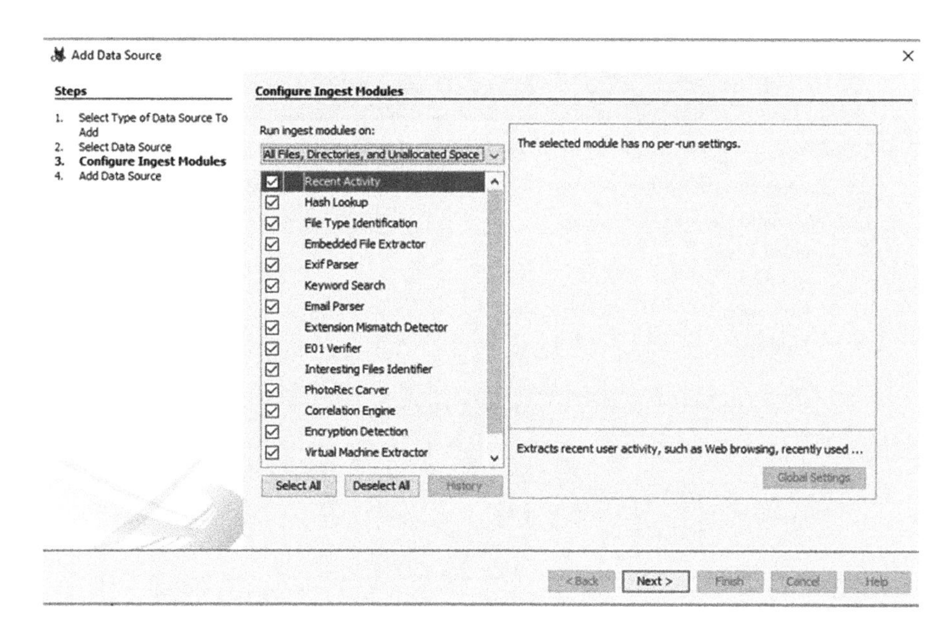

FIGURE 6.15 Autopsy extract data.

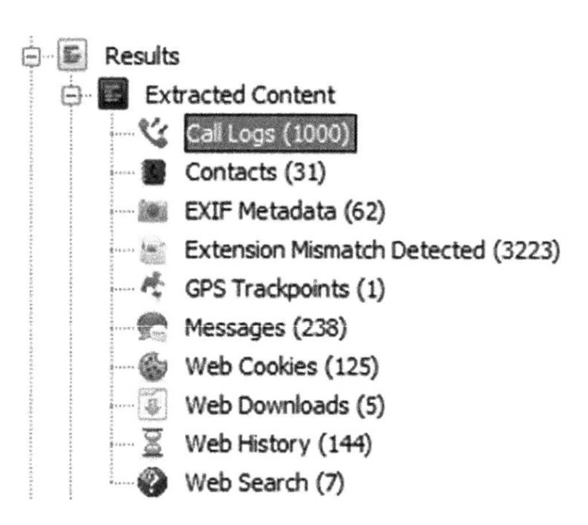

FIGURE 6.16 Autopsy extracted data.

It is beyond the scope of this chapter to provide a detailed coverage of all the tools in the virtual machine. Many of the tools are python scripts, and in some cases simply instructions for how to manually perform tasks such as imaging a phone. You will also notice there are tools for app decompiling and analysis, which we will discuss later in this chapter.

BitPim Settings

Read Only — Block writing anything to the phone
Disk storage — C:\Users\Administrator\Documents\bitpim
Config File — C:\Users\Administrator\Documents\bitpim\.bitpim
Phone Type — SPH-M300PIM — Phone Wizard...
Com Port — auto — Browse ...
Com Timeout (sec) — 3.0
Check for Update — Never
Startup — Always start with the Today tab
Task Bar Icon — Place BitPim Icon in the System Tray when Minimized
Place BitPim Icon in the System Tray when Closed
Autodetect at Startup — Detect phone at bitpim startup
SplashScreen Time (sec) — 2.5
BitFling — Enabled — Settings ...
OK Cancel Help

FIGURE 6.17 BitPim main screen.

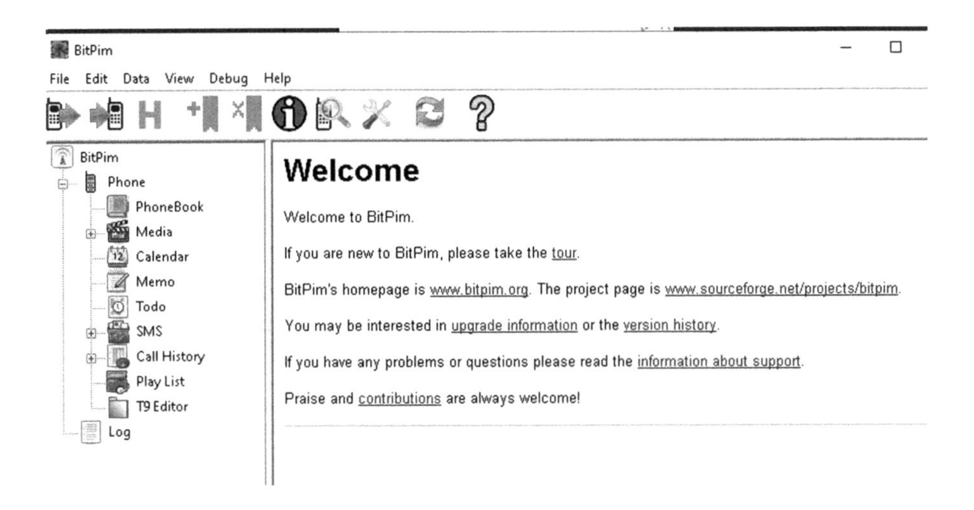

FIGURE 6.18 BitPim help screen.

DUAL USE TOOLS

The following tools work with Android or iOS. These were discussed in Chapter 5. They are covered again here because some readers will want to focus on only those chapters that are relevant to their own investigations and may have skipped Chapter 5.

FIGURE 6.19 OSAF main screen.

Cellebrite

This is probably the most widely known phone forensics tool. It is used heavily by federal law enforcement. It is also well respected in the industry. It is a very robust and effective tool. The only downside to Cellebrite that I am aware of is its high cost. It is the most expensive phone forensics tool I am aware of. Most of the Cellebrite literature won't tell you directly the cost, you need to speak to a salesperson to get a bid. In general, one can expect to spend in the neighborhood of $10,000 for a Cellebrite license. Cellebrite is an Israeli company known not only for their tools, but also for mobile forensics research.

While this tool is quite popular with law enforcement, and well respected, there is not a screenshot-by-screenshot description of the tool in this chapter. This is primarily due to the fact the Cellebrite is not a single tool. There are a number of tools available from Cellebrite including[4]:

- Cellebrite UFED
- Cellebrite Physical Analyzer
- Cellebrite UFED Cloud
- Cellebrite Premium
- Cellebrite Blacklight
- Cellebrite Commander

Even these products have variations. For example, UFED has UFED Touch 2, UFED Touch 2 Ruggedized, and others. It would take an entire book to adequately describe the various Cellebrite products. And, as was discussed earlier, each can be rather expensive. It is also the case that the Cellebrite tools usually require formal training, beyond the scope of this chapter or book. The primary focus in this chapter will be with tools that are more affordable. But more importantly, tools that don't require extensive training.

Dr. Fone

Dr. Fone is a widely used tool for mobile device recovery and transferring of data. This makes it particularly interesting for forensics. This tool is very inexpensive and works with both iPhone and Android. The tool can be found at https://drfone.wondershare.net. The full version is $139.95. The main screen is shown in Figure 6.20.

The information tab will allow you to view SMS messages and phone numbers. These are often critical to a digital forensic investigation. Using Dr. Fone you can see the file system discussed earlier. This is shown in Figure 6.21.

Most important will be the ability to copy data from the phone to your forensics workstation. As you can see, Dr. Fone supports this ability. There is the backup option and the transfer option shown in Figure 6.22.

OXYGEN FORENSICS

Oxygen Forensics is known primarily for its easy-to-use interface. The initial connection with a mobile device is accomplished via a wizard, thus making it quite easy to use. It does not have all of the features one finds in tools such as Cellebrite, but certainly many more than tools like Dr. Fone. The pricing is in the neighborhood of $7000 per license. The company website is https://www.oxygen-forensic.com/en/. They formally offered two tools: the Detective and the Analyst version. Now they only offer the Detective version. The wizard is very easy to use, and the first step of that wizard is shown in Figure 6.23.

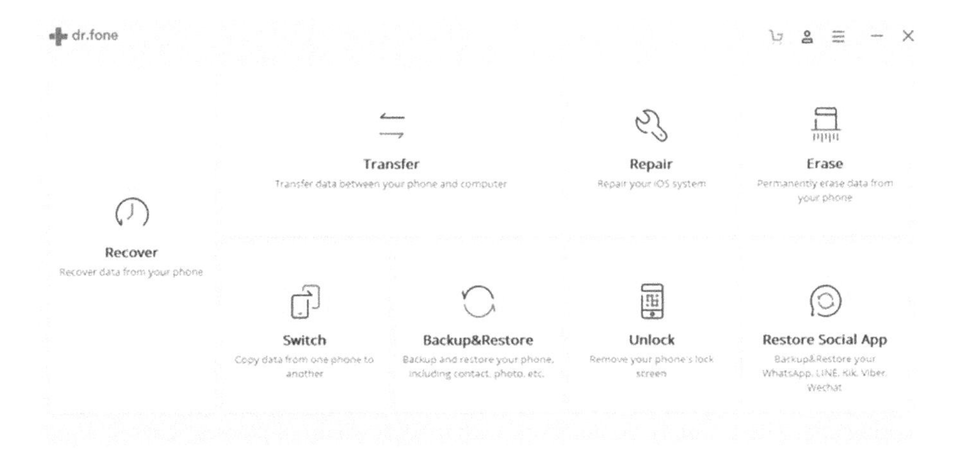

FIGURE 6.20 Dr. Fone main screen.

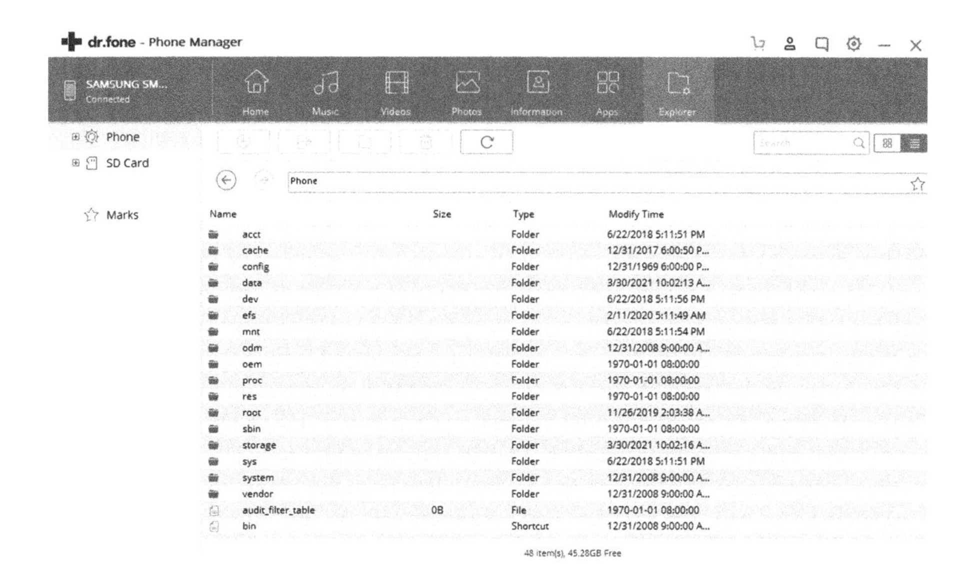

FIGURE 6.21 OSAF main screen. Dr. Fone file system

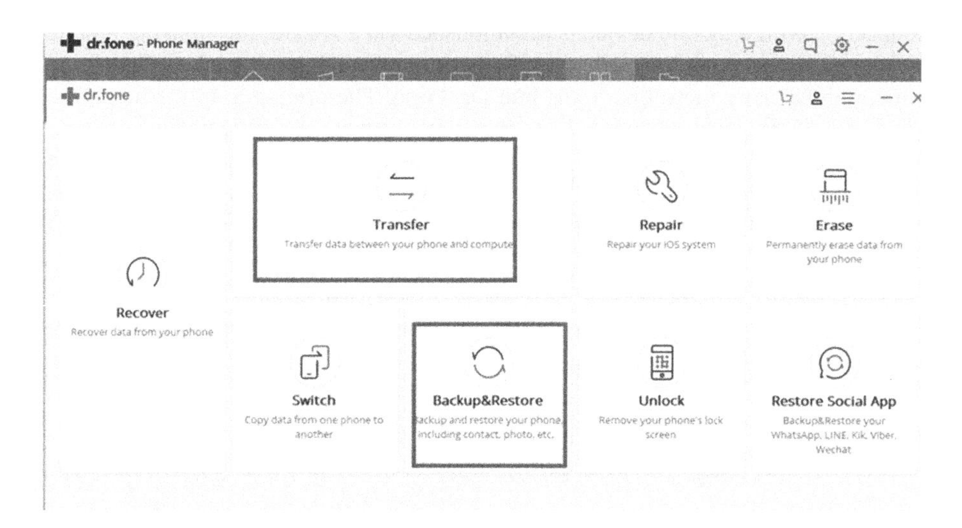

FIGURE 6.22 Dr. Fone transfer data.

As was stated earlier, the main benefit of Oxygen is a rather easy to use interface. Once the extraction is done, the results are very easy to work with. It should be noted that if you try physical access with an Android, that Oxygen will attempt a rooting app. These simply do not work with modern Android phones. Therefore, in the Wizard choose logical access, unless you have already rooted the phone. This is shown in Figure 6.24.

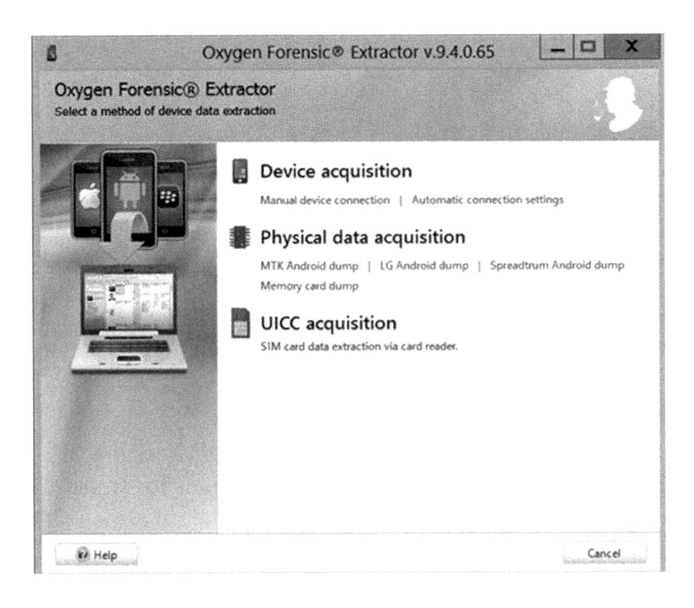

FIGURE 6.23 Oxygen Forensics wizard.

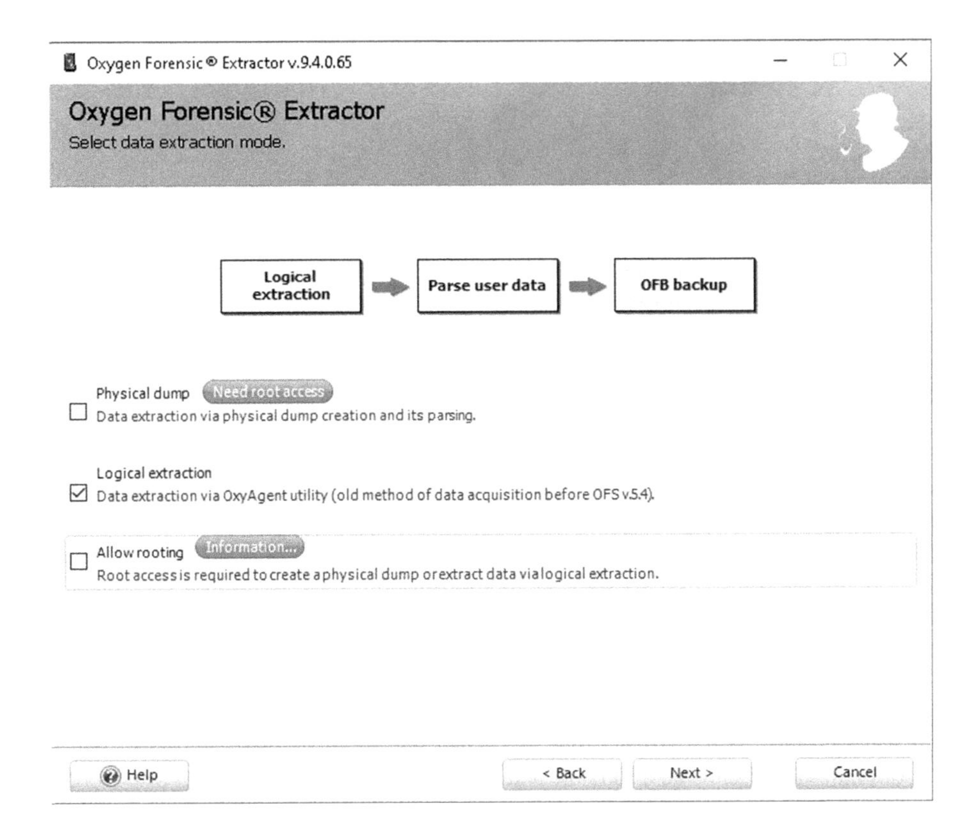

FIGURE 6.24 Oxygen Forensics logical access.

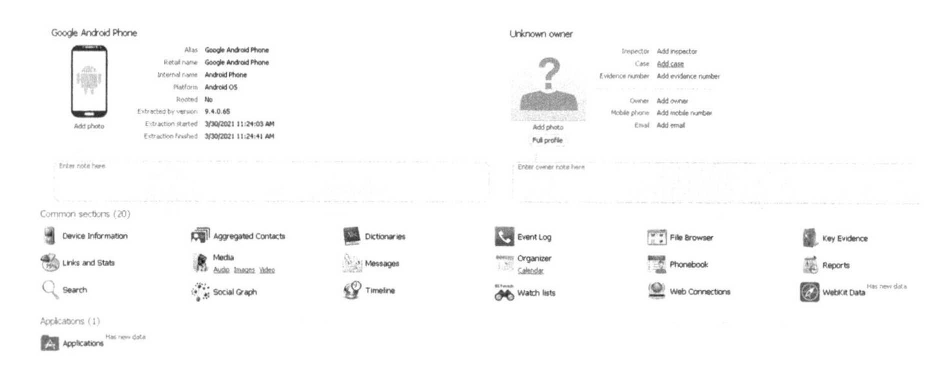

FIGURE 6.25 Oxygen Forensics results.

Figure 6.25 is a screenshot of Oxygen Forensics results for an older phone I use for forensics labs.

As can be seen, there is a substantial amount of data presented in an easy to find interface. It is quite easy to navigate to events, phone books, messages, and many other pieces of data the forensic examiner may have an interest in. In general, Oxygen is a robust tool with a number of interesting features. It is a reasonable option for the professional forensics lab to include. Given the cost of forensics tools, it is recommended that you seek out recommendations from colleagues, and not rely totally on the marketing information from vendors.

MobilEdit

MobilEdit is a low-priced forensics tool that has a number of professional-level features. The MobilEdit Forensic Express is the tool they recommend for the most robust forensics. There are different prices. The Forensic Express Standard is $1500 per license, and you need to contact their sales department to get more information on the pricing of Forensic Express Pro. The company website is https://www.mobiledit.com/forensic-express. The starting screen for MobilEdit is shown in Figure 6.26.

Note that you may need to download drivers specific for your Android model for the connection to work properly. Once the device is recognized by the software, the examiner has an opportunity to enter case details. This is important, as case management becomes rather complex as you have a growing caseload.

Perhaps the most user-friendly aspect of MobilEdit is the various reporting formats. As you can see in Figure 6.27, the examiner can select multiple formats for the report. The HTML report is easy to navigate. However, the PDF report is often easiest to submit to some third party such as an attorney or other party.

The HTML report is shown in Figure 6.28. This format allows one to simply click on the link on the left in order to navigate to a portion of the report.

Overall, MobilEdit is an affordable and reasonably fully featured tool. It also works with both Apple and Android phones, making it useful on most mobile devices.

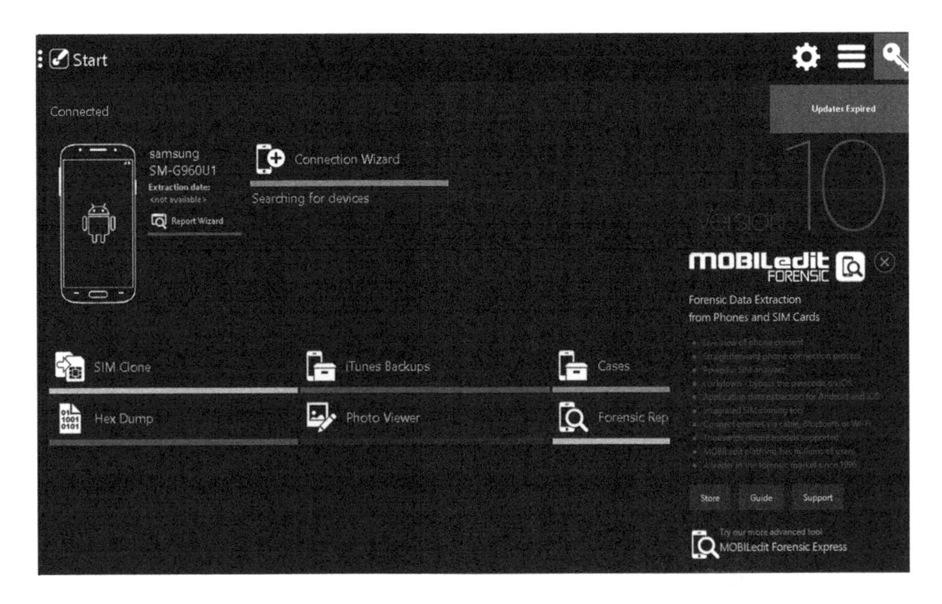

FIGURE 6.26 MobilEdit main screen.

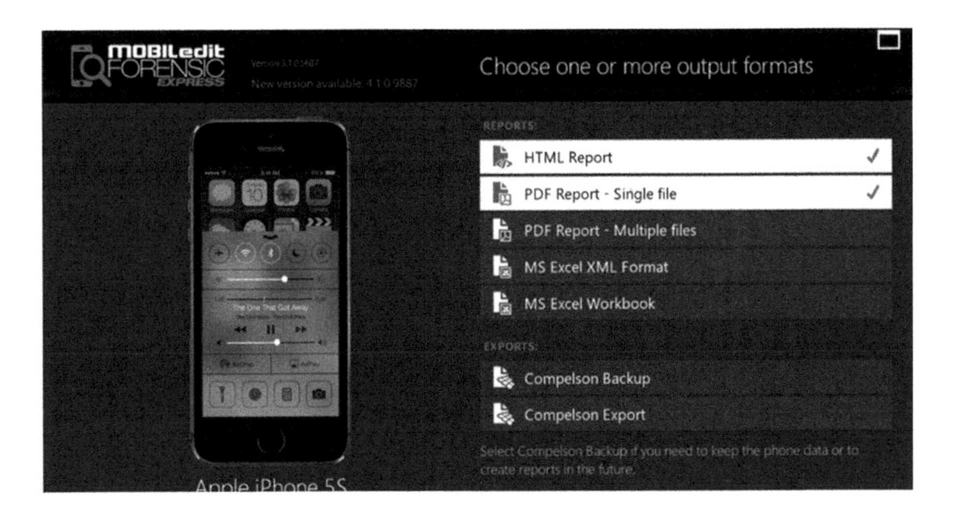

FIGURE 6.27 MobilEdit report options.

ANDROID APP DECOMPILING

Forensics frequently involves understanding the apps on the phone. The apps could be malware for some time. In other instances, someone might claim that malware on their phone is responsible for illegal content, and it is necessary to be able to view the app to determine if this is true or not. It is fortunately quite easy to decompile Android apps.

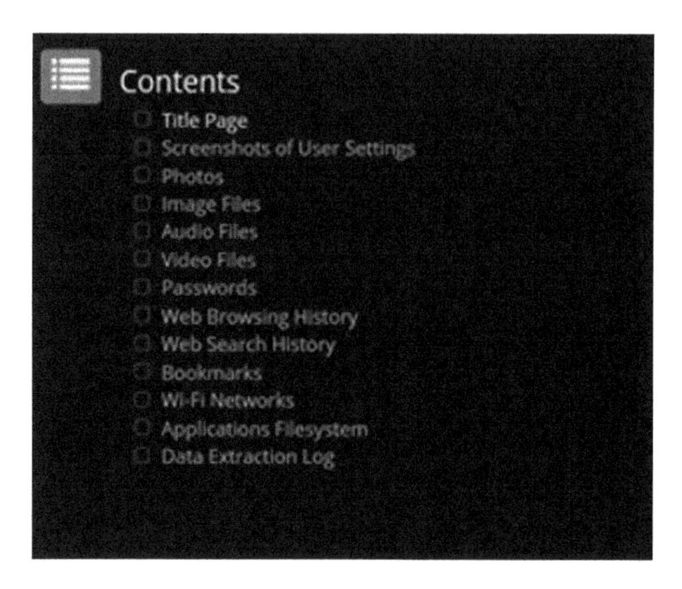

FIGURE 6.28 MobilEdit report example.

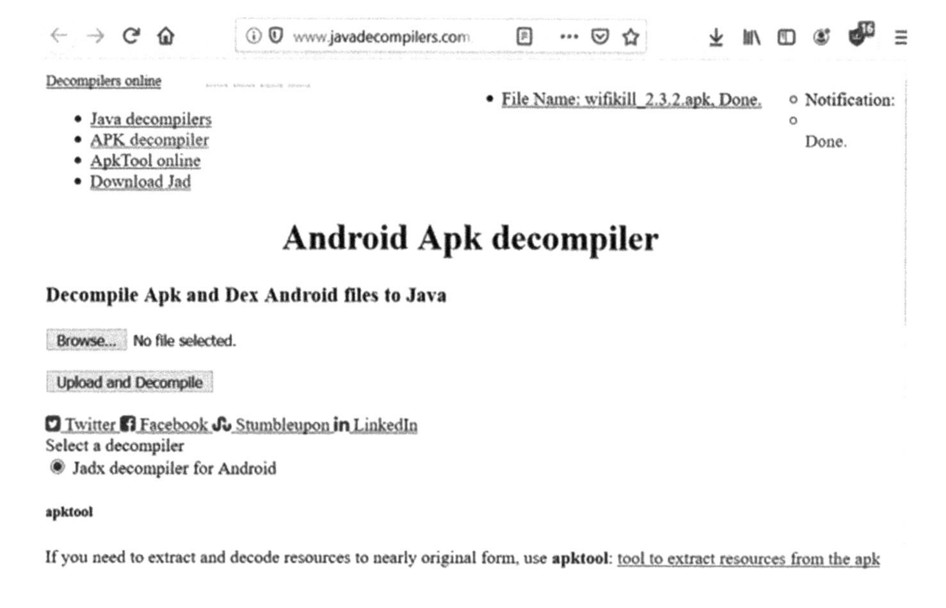

FIGURE 6.29 http://www.javadecompilers.com/apk.

One such online decompiler can be found at http://www.javadecompilers.com/apk. You simply browse to the APK in question then upload it to the website. You can see this in Figure 6.29.

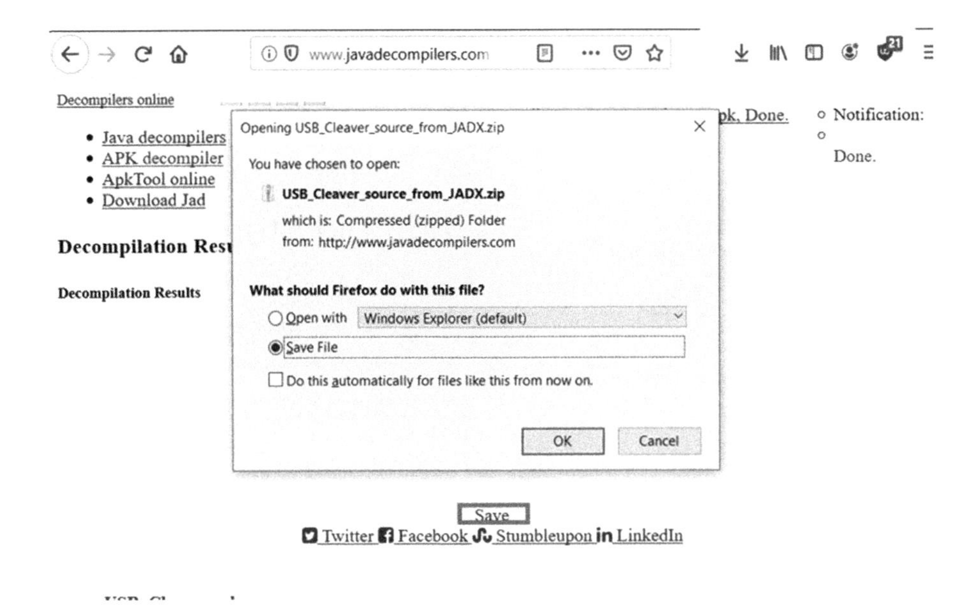

FIGURE 6.30 Download decompiled code.

When the tool is done you will be able to download the source code for that APK. This is shown in Figure 6.30.

It is also possible to use the Android Studio to decompile and debug apps. It is one of the options on the main screen. This is shown in Figure 6.31.

This starts a user-friendly wizard that guides you to select the apk you wish to decompile. You can see this in Figure 6.32.

There are other decompilers you can try online:

- http://www.decompileandroid.com/
- https://www.apkdecompilers.com/

The specific decompiler you use is less important than analyzing the code. Most Android apps are written in either Java or Kotlin programming languages. One need not be an expert programmer in order to follow along with the code. However, at least basic programming skills are required. It is beyond the scope of this book to teach programming. It may be that you will require a consultant who is a programmer to interpret the decompiled code.

CHAPTER SUMMARY

In chapter 6 we have focused on Android forensics. Combining this information with the information from chapter 4 should give you a strong working knowledge of the Android operating system and how to perform forensics. Specifically, we have

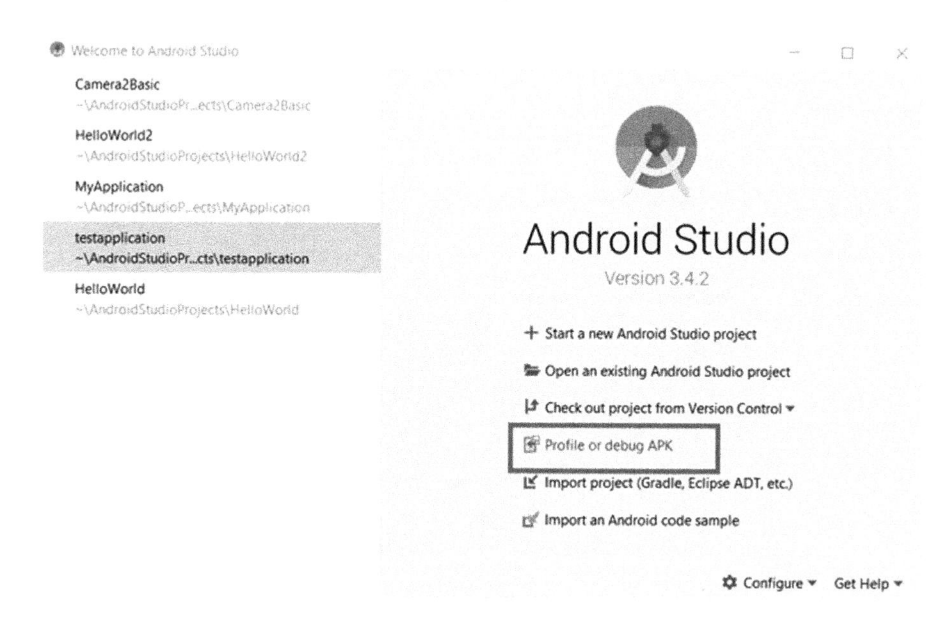

FIGURE 6.31 Decompile with Android Studio.

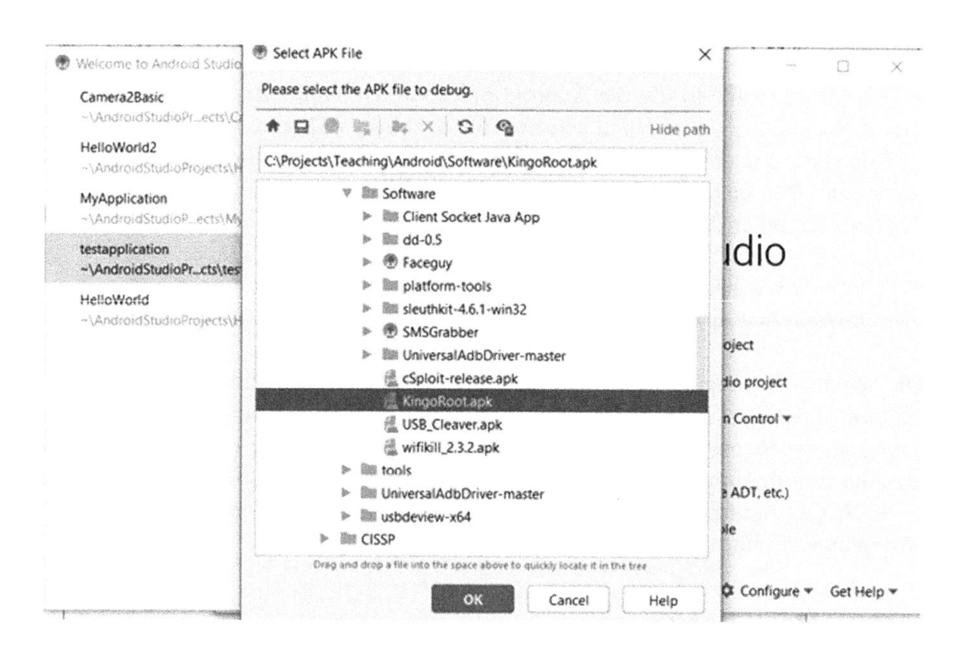

FIGURE 6.32 Selecting an APK.

covered a range of tools. Some of those are Android specific, and some work for different mobile devices. It is important that you have a working knowledge of at least some of these tools.

CHAPTER 6 ASSESSMENT

1. The _____ is a unique address that identifies the access point/router that creates the wireless network.
   a. MAC Address
   b. SSID
   c. BSSID
   d. WAP Name
2. According to SWGDE what is the 2nd tier in the forensics pyramid?
   a. Manual
   b. Logical
   c. Physical
   d. JTAG
3. What is the best way to describe what must be in an expert report?
   a. The tools used.
   b. The techniques used.
   c. The process used.
   d. The basis and reasons for all opinions.
4. How many communication apps can most forensics tools acquire data from?
   a. Less than six
   b. All apps
   c. A few
   d. Most
5. What, if anything, is incorrect about this statement: adb backup -all -f
   a. Nothing
   b. No file path
   c. No source path
   d. Should be from shell

CHAPTER 6 LABS

For these labs you will need an Android phone. Even an older model, or one with defects such as a broken screen will be adequate. As long as it powers on, that is the issue. Such phones can be found used on E-Bay, Amazon.com, and phone repair stores for approximately $20 US Dollars.

Lab 6.1 ADB

First backup your phone using ADB

adb backup -all -f C: \backup.ab

Next connect to the device starting with

adb devices

You may need to troubleshoot issues here.

Then try adb shell

This will be followed by navigating the phones file directory using

```
ls
ls -f
ps
netstat
dumpstate
getprop ro.product.model
getprop ro.build.version.release
getprop ro.serialno
getprop ro.product.name
getprop ro.product.cpu.abi
getprop ro.build.fingerprint
getprop ro.product.locale.language
getprop ro.wifi.channels
getprop gsm.baseband.imei
getprop ro.build.date
```

You should write a lab report as if this were an actual case. Describe your processes in detail, what you found, any issues you encountered. Consider the guidelines for forensics reports discussed in this chapter.

Lab 6.2 Android Tools

Download Android Tools from https://sourceforge.net/projects/android-tools/.

Use this to extract similar data from the phone, as you did in lab 1.

You should write a lab report as if this were an actual case. Describe your processes in detail, what you found, any issues you encountered. Consider the guidelines for forensics reports discussed in this chapter.

NOTES

1 https://www.ncjrs.gov/pdffiles1/nij/199408.pdf
2 https://www.sans.org/blog/intro-to-report-writing-for-digital-forensics/
3 National Research Council. (2011). *Reference manual on scientific evidence*. National Academies Press.
4 https://www.cellebrite.com/en/product/

7 JTAG and Chip-Off

Traditional forensics means sometimes simply cannot get the evidence. In those situations you may need to try more in depth technical data extraction. The two most common methods to dig deeper, at least for Android phones, are JTAG and Chip-off. Each of these methods use electrical engineering test techniques to extract data from the phone by accessing the chips directly.

JTAG and chip-off are two methods for physically accessing the data on chips in the phone. The phone's operating system is bypassed, and the examiner goes directly to the actual circuitry. The term JTAG is an acronym for Joint Test Action Group. The techniques used were designed for testing printed circuit boards. It was later discovered that these same techniques could perhaps gather forensic evidence.

JTAG is a forensic technique is often sought because it bypasses the operating system. That means it bypasses all operating system security. That includes PIN, password, biometrics, patterns, etc. JTAG does not work on iOS devices, but has been successful on Android devices. The process can be tedious and slow, but in some cases may be the only way to get at the data on the device.

In 1985 the Joint Test Action Group (JTAG) was formed to provide methods for testing chips. In 1990, JTAG became an IEEE (Institute of Electrical and Electronics Engineers) standard, standard 1149.1–1990. This standard was supplemented in 1994 with the boundary scan description language (BSDL). Boundary scanning is the key to the forensic application of JTAG. BSDL defines the signal for each pin on a chip. The JTAG process allows the engineer to access data internal to the device, this enables testing of the chip. The following quotation, while lengthy, should help to explain JTAG:

> Circuitry that may be built into an integrated circuit to assist in the test, maintenance and support of assembled printed circuit boards and the test of internal circuits is defined. The circuitry includes a standard interface through which instructions and test data are communicated. A set of test features is defined, including a boundary-scan register, such that the component is able Circuitry that may be built into an integrated circuit to assist in the test, maintenance and support of assembled printed circuit boards and the test of internal circuits is defined. The circuitry includes a standard interface through which instructions and test data are communicated. A set of test features is defined, including a boundary-scan register, such that the component is able to respond to a minimum set of instructions designed to assist with testing of assembled printed circuit boards. Also, a language is defined that allows rigorous structural description of the component-specific aspects of such testability features, and a second language is defined that allows rigorous procedural description of how the testability features may be used.

DOI: 10.1201/9781003118718-7

-1149.1-2013 – IEEE Standard for Test Access Port and Boundary-Scan Architecture

The key to working with JTAG is the test access points (TAPs). The engineer manipulates these to conduct testing. With forensics, the TAPS are manipulated to retrieve data directly from the chip. This is why JTAG became of interest in the forensic community. It bypasses the operating system. So, any screen locks on the phone can be bypassed. However, if the data is encrypted JTAG will not be useful. Any data retrieved will still be encrypted.

ELECTRONICS 101

In order to understand both JTAG and chip-off, the reader will need a basic understanding of electronics. This section provides a very basic understanding. Electronics is the use of electrical signals to accomplish some goal. Electronics can be used to provide lighting, temperature control, and a range of other functions. However, for the purposes of this book, the primary function of concern is the storage and transmission of data.

Let us begin with explaining some concepts and terms. First we have electric charge. Charge is a term for electrical energy. It is often measured in coulombs. A coulomb is approximately 6.242×10^{18} electrons. An ampere is defined as 1 coulomb moving past a point in one second. This provides us with an understanding of both charge and current. The charge being measured in coulombs and the current in amperes.

The flow of electrical energy is described as either direct current (DC) or alternating current (AC). The term current describes the flow of electrons, which is the actual form of electrical energy. DC means that the current is in one steady direction. One of the most common sources of direct current is a battery. Electrical energy from a battery flows in a steady direction from the negative (–) terminal to the positive (+) terminal. AC does not flow in one direction. It alternates from a negative charge to a positive charge. The most common source of alternating current is the wall outlet.

Resistance is the opposition to direct current (DC). Every conductor has a certain amount of resistance that affects current – the longer the conductor, the greater the resistance. Resistance is a DC term and should not be confused with impedance, which refers to the opposition in alternating current (AC). Impedance increases as frequency increases. In other words, the higher the frequency, the faster the speed of data transmission. However, the quality or integrity of the data will diminish because of impedance. This is why there are limitations to the speed at which data can be delivered across a conductor. The length of a conductor also influences impedance. As the length of a conductor increases, so does the total impedance.

Electrical power is often measured in watts or megawatts. A watt is the equivalent of 1 joule of energy per second. A joule is the among of work done by a force of one newton moving an object one meter. And a newton is defined as the amount of force required to accelerate a mass of one kilogram at a rate of one meter per second, per second. Now we have a basic definition of current, charge, and power.

Next let us define some basic electronic components. An inductor is a component that induces an electric charge. A capacitor stores an electric charge. Resistors are used to reduce the flow of current. The resister may seem a bit odd. But consider a circuit board that has current flowing at some particular value. Then consider a component that requires a little less current. Placing a resister before that component solves the problem. There are some essential terms you will need; the basic terminology of electronics is defined in this list:

Alternating Current – Direction reverses
Current requires some source and a sink. The difference between source and sink is the potential.
Diode – a two terminal component that conducts current in one direction.
Direct Current – Direction does not reverse.
Discrete Device – a traditional electronic device, e.g., resistor, capacitor, battery
Electromotive Force (EMF) – Force that causes electrons to move in an electric field. The volt is the unit of electromagnetic force.
Farad is the unit of capacitance. Usually measured in milli-farads or mF.
Integrated Circuit (IC) – a combination of discrete devices that may support some specific functionality.
Package – the container in which an IC or discrete device is implemented.
Printed Circuit Board – a board etched with circuitry (usually copper) to create electronic circuits.
Surface Mounted Device – a package for discrete components/IC that allows direct soldering to the board.
Through Hole Technology – Circuits are built with components that are soldered through holes.
Voltage can store binary data; low voltages is a 0 high voltage is a 1.

Electric power is often described by Ohm's law. This law states that the current between two points is proportional to the voltage across the two points. This is expressed in the formula seen in Equation 7.1.

$$I = \frac{v}{R} \tag{7.1}$$

In Equation 7.1, I is the current, v is the voltage, and R is the resistance. The unit of resistance is the ohm, symbolized by Ω. There are symbols used in electronics that you should be familiar with. These are shown in Table 7.1. Is only the most basic symbols, there are many more. However, these symbols are enough for you to understand diagrams related to the JTAG process.

BOUNDARY SCANNING

Boundary-Scan Description Language (BSDL) files are used to describe the boundary-scan behavior and capabilities of a given device. The BSDL describes important properties of a given device's boundary-scan functions, including:

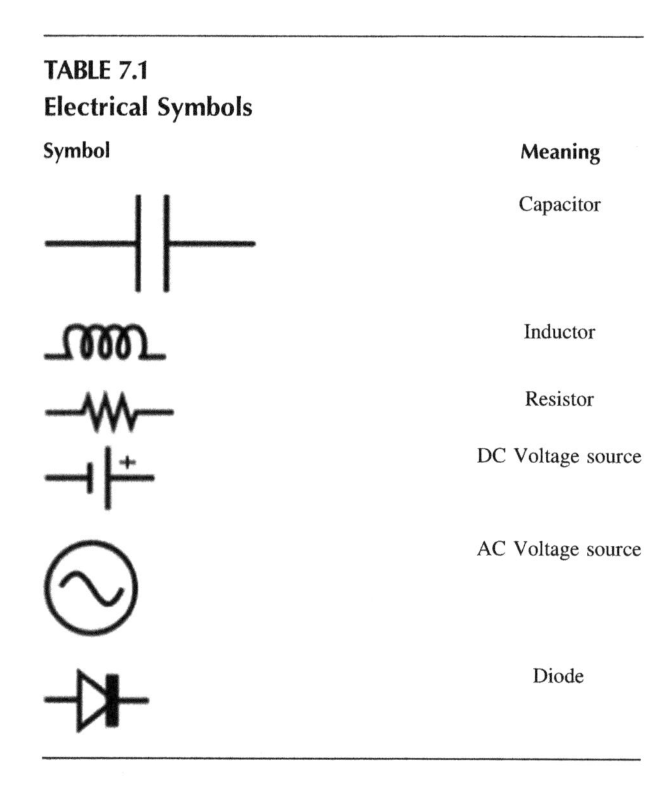

TABLE 7.1
Electrical Symbols

Symbol	Meaning
	Capacitor
	Inductor
	Resistor
+	DC Voltage source
	AC Voltage source
	Diode

- Which JTAG standards are supported by the device
- Signal mapping and package information
- Available instructions, and which registers those instructions access
- The type of boundary-scan cell available for each signal
- Information about signals that affect compliance to the standard
- Design warnings and notes

Boundary scanning is what JTAG forensics is using. Essentially one connects to the various test access points and sends a signal into them. This signal will run through the chips and allow you to extract the data on the chips.

JTAG Terms

Previously we explored basic electronics terminology which you will need to understand JTAG. There are also JTAG specific terms you will need to be familiar with in order to understand the processes of JTAG and Chip-off.

JTAG = Joint Test Action Group – method for testing circuits
IEEE Standard 1149.1 – JTAG standard
TAP – Test Access Port
DCC – Debug Communications Channel
BSR – Boundary Scan Register

TDI (Test Data In)
TDO (Test Data Out)
TCK (Test Clock)
TMS (Test Mode Select)
TRST (Test Reset) optional
ETM – Embedded Trace Module
DSCR – Debug Status and Control Register
ETB – Embedded Trace Buffer
ITR – Instruction Transfer Register

Each of these terms is important to the JTAG process. However, the Test terms (i.e., test data in, test data out, test clock, etc.) are the actual names of test access ports (TAPs) that you will connect to in order to perform JTAG.

The signals are represented in the boundary scan register (BSR) accessible via the TAP.

IEEE 1149.1 defines these connector pins:

- **TDI** (Test Data In)
- **TDO** (Test Data Out)
- **TCK** (Test Clock)
- **TMS** (Test Mode Select)
- **TRST** (Test Reset) optional

One of the most challenging aspects of JTAG is finding the TAPs. They are not all in the same location on every device. Furthermore, there is not a single comprehensive reference for the TAPs for the various models of Android phone. Figure 7.1 shows the TAPs for one specific phone model.

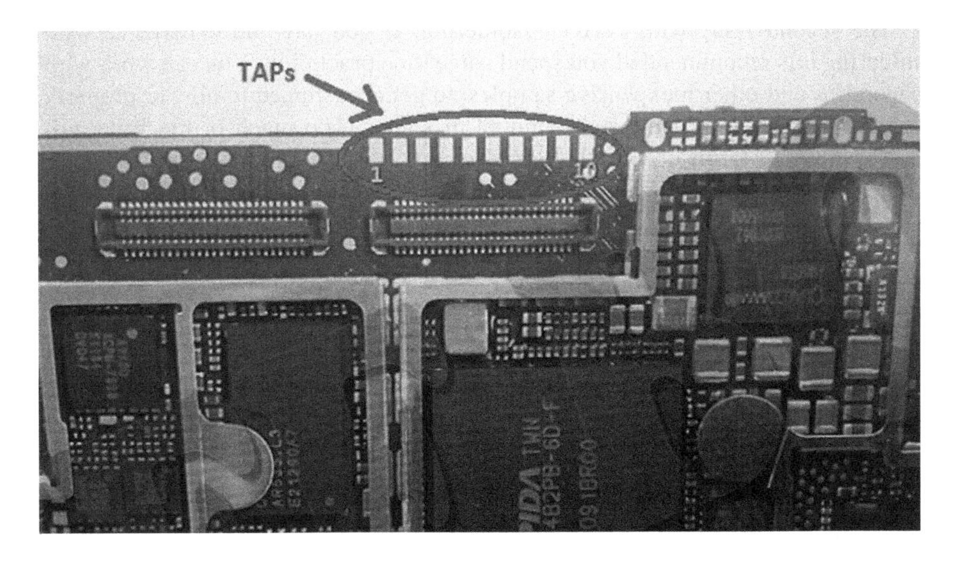

FIGURE 7.1 Phone TAPs Example 1.

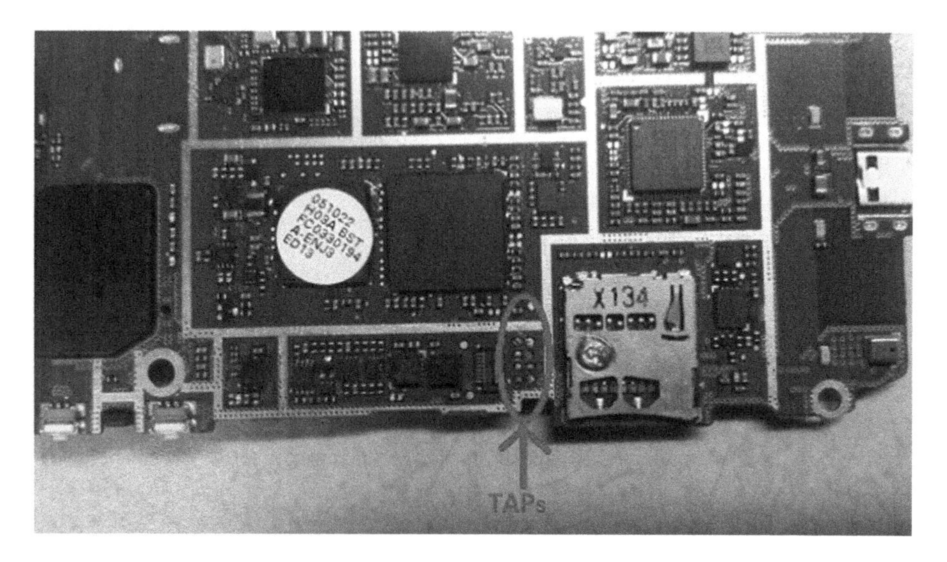

FIGURE 7.2 Phone TAPs Example 2.

These are relatively easy to find, and even numbered 1 to 10. However, the first issue is that it is not clear which numbered TAP is TDI, TCO, TCK, etc. The next issue is that the TAPs can look entirely different on another device. You can see another phone's TAPs in Figure 7.2.

The best advice is to search the internet for documentation on the TAPS for a given model of phone. There is no guarantee you will be able to find the TAPs for the model phone you are working with. As you learn more TAP layouts for various models it is advised that you keep a record of the TAPs. You may need them later.

The second issue with JTAG is soldering. If you have no experience with soldering it is recommended you spend some time practicing. You can work with paperclips and other inexpensive samples, to get accustomed to how to properly solder. The key is just the right amount of solder. Too much or too little will make your JTAG unsuccessful. It is also important to use the correct type of solder. .032-gauge solder is recommended that is either lead free or 60% tin /40% lead.

A third issue regards battery. Some phones won't work if they cannot see the battery recharge. This can occur in several Nokia and Motorola brands. To fix this just do the following steps:

1. Remove the battery
2. Attach wires to the battery contacts
3. Place battery in the phone so that the wires are "sandwiched" between the phone and the battery
4. Attach the power supply to the exposed wires
5. Phone will see the battery charge sensing pin and the power supply will be able to power the phone.

Common JTAG Tools

There are a number of tools available to assist you in JTAG operations. Each of these has its strengths and weaknesses. It is not the goal of this chapter or this book to endorse any particular tool. However, it is advantageous for you as a forensic examiner to be familiar with a range of tools.

RIFF

RIFF Box has been in the JTAG business for many years. The tool is widely used and respected. The company website is https://www.riffbox.org/. On the website you will find documentation including the software used with the RIFF box. They also have a JTAG getting started guide found at https://www.riffbox.org/downloads/GettingStarted.pdf. The RIFF box is shown in Figure 7.3.

The RIFF box software is relatively easy to use and is shown in Figure 7.4.

Some common RIFF errors and their causes/solutions are described here:

Connecting to the dead body…ERROR

1. ERROR: The RTCK Signal does not respond. The cause is often that the target device is hardware damaged – check target device hardware for faults.
2. ERROR: Target device is not powered on – check resurrection help for instructions how to power device.
3. ERROR: Connection between target and box is not done properly. Common causes are bad cables or bad pinout. The solution is to rework or replace 20 pin ribbon cable, and check pinout and soldering.
4. ERROR: RIFF Box hardware not working. This requires you to first test with 100% working phone, if doesn't work, replace box or send for repair.
5. ERROR: DCC Loader Execute failed. Terminating. This has a more involved solution process:

FIGURE 7.3 RIFF box.

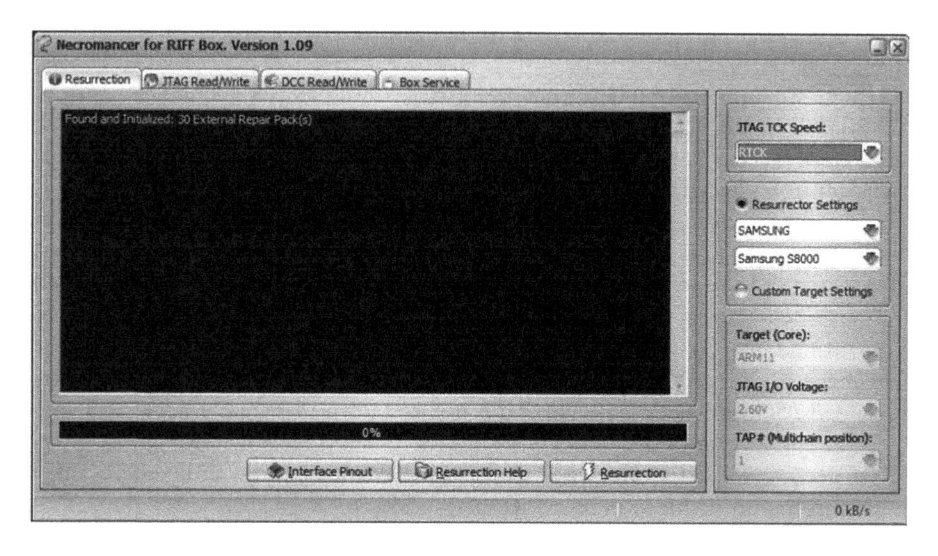

FIGURE 7.4 RIFF software.

a. Check if NRST signal is connected.
b. Change RTCK Sampling from "Sample At Max" to lower value, for example 1 MHz. Not all devices will work at maximum speed.
c. Check target power and connection.

6. ERROR: The RTCK Signal does not respond. This problem can have several different causes:
   a. RTCK not connected.
   b. RTCK signal doesn't exist for current target device.
   c. Target device is connected improperly.
   d. Target device is not powered on.
   e. Target device hardware is malfunctioning.
7. As you might surmise, having multiple possible causes means there are multiple possible solutions:
   a. If target device pinout contains RTCK signal pin, connect it.
   b. If target device doesn't contain RTCK signal pin, use fixed TCK value.
   c. Recheck connection according to provided JTAG target device pinout.
   d. Make sure that target device has sufficient power and that power on sequence is executed.

ORT Box

ORT (Omni Repair Tool) is available from https://www.ort-jtag.com/. Their downloads are only available to those who had purchased and registered the ORT tool. The ORT JTAG box is shown in Figure 7.5.

FIGURE 7.5 ORT box.

Easy JTAG Box

Easy JTAG is available from https://easy-jtag.com/. This tool is interesting because it has specific add-ons for Samsung devices. Their tool is widely used and respected. The box can be seen in Figure 7.6.

FIGURE 7.6 Easy JTAG box.

JTAG STANDARDS

As with other areas of mobile device forensics, there are standards for the JTAG process. We will examine a few of these standards briefly in this section.

SWGDE

The Scientific Working Group on Digital Evidence has standards that are available at https://www.irisinvestigations.com/wp-content/uploads/2019/05/SWGDE-Best-Practices-for-Examining-Mobile-Phones-Using-JTAG-092915.pdf. What is most interesting about the SWGDE standard is that it not only describes the JTAG process, but what should be included in JTAG training.

IEEE 1149

IEEE 1149 is the standard that defines JTAG. This standard does not describe how to perform JTAG forensics. Rather it defines the entire JTAG engineering process. A good place to begin will be https://www.jtag.com/ieee-1149-1/. This standard won't be quite as directly helpful for forensics but will give you a good engineering understanding of JTAG.

NIST

As was mentioned earlier in this book, the U.S. National Institute of Standards defines computer forensic tool testing standards. This also applies to JTAG tools. There are several links to NIST information on JTAG:

https://www.nist.gov/news-events/news/2020/01/nist-tests-forensic-methods-getting-data-damaged-mobile-phones
https://www.nist.gov/image/20itl003jtagmethodjpg
https://www.nist.gov/system/files/documents/2020/08/21/JTAG%20and%20Chip-Off%20Data%20Analysis%20and%20Testing_AAFS_2020.pdf

JTAG Guidance

While not actually a standard, the District of Columbia provides guidance on JTAG forensics. The document can be found at https://dfs.dc.gov/sites/default/files/dc/sites/dfs/publication/attachments/DEUSOP09%20-%20Using%20JTAG-ISP%20for%20Mobile%20Device%20Examinations.pdf. Similarly, the Department of Homeland Security has a document discussing JTAG and Chip-Off data extraction that is publicly available at https://www.dhs.gov/sites/default/files/publications/testresultsforbinaryimage-decodingandanalysistool-x-ways_v19.8_sr-7.pdf

CHIP-OFF

Chip-off could be thought of as an extension to JTAG. Rather than attaching wires to TAPs, the phones chip is unsoldered and placed in a chip reader.

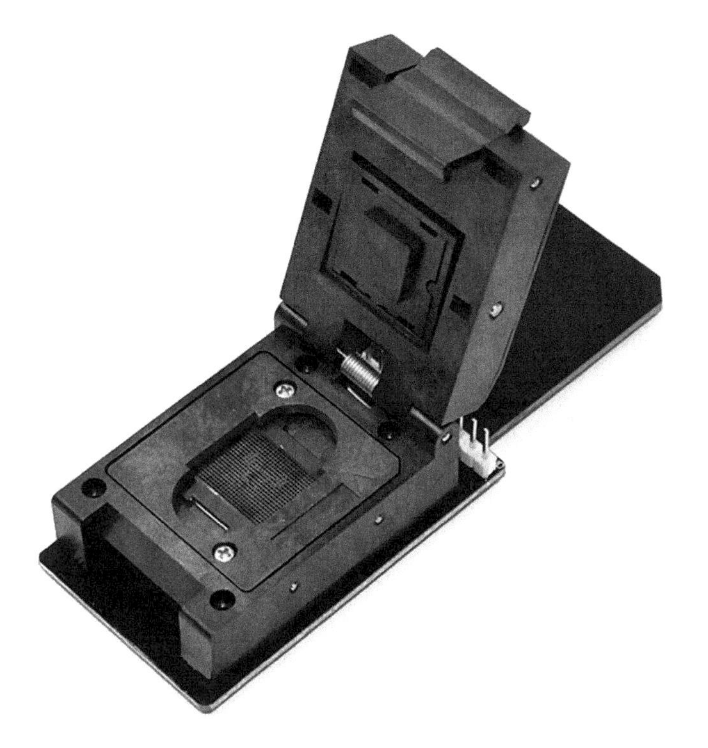

FIGURE 7.7 Chip reader.

Fortunately, many mobile devices use a TSOP style NAND chip. TSOP refers to Thin Small-Outline Package. NAND refers to how the chip performs logical operations, in this case NAND means Not AND. This is helpful given that many phones will have the same general chip design. A standard chip reader is shown in Figure 7.7.

The primary difficulty is in getting the chip unsoldered. If that is done successfully, without damaging the chip, then some aspects of chip-off are actually easier than JTAG. Rather than try to find TAPs and match them to wires, you simply place the chip into the chip reader. The reader is then connected to a forensic workstation, and software will read the data from the chip. It should be noted that chip-off is a destructive process.

CHAPTER SUMMARY

In this chapter we briefly examined two different forensic techniques. The techniques are JTAG and chip-ff. Both of these techniques are methods for going directly to the device computer chip to get evidence. Both require some level of basic electronic skill. Either method is effective at circumventing Android operating system forensics, however, neither method can access encrypted data.

CHAPTER 7 ASSESSMENT

1. An ampere is a unit of what?
   a. Charge
   b. Current
   c. Resistance
   d. Power
2. A two terminal component that conducts current in one direction is called a _____
   a. diode
   b. discrete device
   c. package
   d. printed circuit board
3. IEEE 1149.1 defines how many TAPs for boundary scanning?
   a. 3
   b. 4
   c. 5
   d. 6
4. (True/False) Chip-off is not as destructive as JTAG.
5. (True/False) TRST is optional.

8 SQLite Database Forensics

The apps running on your phone, be it an Android or iOS device, store data in databases. The most common database used in phones is SQLite. Most major forensics tools will be able to extract data for some apps, but not all. There will be times when you will need to extract data directly from the SQLite database.

Both Android and iPhone make use of SQLite databases. These are basic relational databases that store data. The various forensics tools you have seen earlier in this book often access these databases to get data such as call logs and messages. As was discussed at some length earlier in this book, no forensic tool can extract from all apps. Therefore, it is important to understand how to access SQLite databases manually and extract data. This will extend your capabilities beyond what is allowed by your tool of choice.

RELATIONAL DATABASES

SQLite is a relational database. This necessitates some discussion of relational databases. Relational databases are based on relations between various tables. The structure includes tables, primary and foreign keys, and relations. A basic description can be summarized with the following points:

- Each row represents a single entity.
- Each column represents a single attribute.
- Each record is identified by a unique number called a *primary key*.
- Tables are related by foreign keys. A *foreign key* is a primary key in another table.

Two sample tables are shown in Figure 8.1. The first table simply lists a foreign key for the column office. That key is the primary key from the office table. This way, if more than one employee is in the same office, the same office information is not repeated. This is one of the strengths of a relational database, eliminating duplicate data.

All relational databases use Structured Query Language (SQL). SQL uses commands such as SELECT, UPDATE, DELETE, INSERT, WHERE, and others. At least the basic queries are very easy to understand and interpret.

STRUCTURED QUERY LANGUAGE

As was stated earlier, all relational databases use SQL. When working with SQLite databases manually, you will need to be able to execute some basic SQL statements in order to extract the data you need. The forensics software you use will typically use SQL

DOI: 10.1201/9781003118718-8

PK	LNAME	FNAME	Office	Hire Date
1	Smith	Jane	2	1/10/2010
2	Perez	Juan	2	1/14/2011
3	Kent	Clark	1	3/2/2005
4	Euler	Leonard	3	3/5/2009
5	Plank	Max	3	4/2/2012

PK	Office City	Office Address	Office Main Phone	Office Manager
1	Manhattan	111 Madison Avenue	555-555-1111	John Smith
2	Los Angeles	222 Ventura Blvd	555-555-2222	Jane Smith
3	Chicago	333 Main Street	555-555-3333	Juan Perez
4	Detroit	444 Maple Avenue	555-555-4444	Juanita Perez

FIGURE 8.1 Relations between tables.

statements on the SQLite databases it is familiar with to extract the data the tool shows you. However, given there are so many different SQLite databases, no forensics tool can retrieve them all. Thus, it is important for you to be able to extract data manually.

Dr. EF Codd published the paper, "A Relational Model of Data for Large Shared Data Banks", in June 1970 in the Association of Computer Machinery (ACM) journal, Communications of the ACM. The language, Structured English Query Language (SEQUEL) was developed by IBM Corporation, Inc., to use Codd's model. SEQUEL later became SQL (still pronounced "sequel"). In 1979, Relational Software, Inc. (now Oracle) introduced the first commercially available implementation of SQL. Today, SQL is accepted as the standard RDBMS SQL Standard. ANSI (American National Standards Institute) and the ISO (International Standards Organization) /IEC (International Electro Technical Commission) have accepted SQL as the standard language for relational databases. The latest SQL standard was adopted in July 1999 and is often called SQL:99.

Fortunately, SQL statements are relatively easy to learn. They appear very much like standard English. Certainly, there are more complex SQL queries, but the essentials are relatively easy to master. In this section we will examine various SQL basic commands providing both the syntax and an example.

The **SQL SELECT** statement is used to select data from a SQL database table. This is usually the very first SQL command students learn and is one of the most used SQL commands.

```
SELECT column_name(s) FROM table_name
Or to get everything in that table:
SELECT * FROM table_name
```

Example
SELECT LastName, FirstName FROM Employees

Sometimes you don't want all of the records. You only want those records that meet some specific criteria. Limiting records to just those meeting a specific criteria is quite easy in SQL. The **SQL WHERE** clause is used to select data conditionally, by adding it to already existing SQL SELECT query

SELECT column_name(s) FROM table_name WHERE column_name operator value
SELECT * FROM Employees WHERE City='Richardson'

Notice that in the example we used the = sign. Fortunately, SQL provides many more options than simply =. Sometimes you want those entries that are not equal, or greater than. Here are some symbols you can use instead of =.

!= Not equal
> Greater than
< Less than
>= Greater than or equal
<= Less than or equal

Just as interesting you can also use queries for items that are similar to what you are asking. Perhaps you don't know the last name, or it could be spelled a few different ways. You would then use the LIKE command. Here is an example:

SELECT * FROM Employees WHERE City LIKE 'Rich%'

This will return any employees residing in any city that starts with "Rich" including Richardson, Richmond, Richland, etc. This can be a quite useful modifier for your query. You may also have more than one criteria. You can use the operators AND/ OR to specific multiple criteria. Here is an example of each:

SELECT * FROM Employees WHERE City='Plano' AND LastName='Johnson'
SELECT * FROM Employees WHERE City='Plano' OR City='Frisco'

You may also require records that are between values. When you use numeric data, the 'between' is rather obvious. When you use text data, SQL views the alphabetical order as the basis for establishing what is between.

SELECT column_name(s) FROM table_name WHERE column_name BETWEEN value1 AND value2
SELECT * FROM Employees WHERE LastName BETWEEN 'Jones' AND 'Smith'

You should also note that not all column names will be a single word such as LastName. If the column name is two words, such as Last Name, you cannot simply enter them in your SQL query in that form. It will generate an error. Instead, you should put them in brackets such as [Last Name].

There will be times when you will want a single record. There may be multiple instances of a given value in the database, but you are seeking a particular record. The DISTINCT qualifier is helpful in this regard. Here is an example:

```
SELECT DISTINCT column_name(s) FROM table_name
Example
SELECT DISTINCT City FROM Employees
```

Regardless of the records you retrieve, you may sometimes need them in some particular order. The ORDER BY qualifier is helpful in this regard. Here you see the general syntax as a couple of examples.

```
SELECT column_name(s) FROM table_name ORDER BY column_name(s)
ASC|DESC
Example
SELECT * FROM Employees ORDER BY LastName
SELECT * FROM Employees ORDER BY LastName ASC
```

By default, order by sorts in descending order. There are certainly more SQL statements, but these should provide you the working knowledge needed to extract data from an SQLite database. You should note that SQL itself is not case sensitive. The capitalization used in this chapter is simply to make the SQL statements more readable. Actual SQL terms are capitalized, other words use sentence case or lower case.

SQLITE BASICS

Both iPhone and Android make use of SQLite databases. These normally end in a .db extension. SQLite is not a client–server database engine. Rather, it is embedded into the end program. SQLite stores the entire database (definitions, tables, indices, and the data itself) as a single cross-platform file on a device, such as a phone. It implements this simple design by locking the entire database file during writing. SQLite read operations can be multitasked, though writes can only be performed sequentially. D Richard Hipp designed SQLite in the spring of 2000 while working for General Dynamics on contract with the United States Navy.

The SQLite data types are:

- NULL – null value
- INTEGER – signed integer, stored in 1, 2, 3, 4, 6, or 8 bytes depending on the magnitude of the value.
- REAL – a floating point value, 8-byte IEEE floating point number.
- TEXT – text string, stored using the database encoding (UTF-8, UTF-16BE or UTF-16LE).
- BLOB. The value is a blob of data, stored exactly as it was input.

If you are able to find SQLite databases on a phone, and can copy them to your forensics machine, you will need an SQLite viewer. You may download an SQLite browser for free from https://sqlitebrowser.org/ This is shown in Figure 8.2.

You open a database by going to File then Open Database and navigating to the database you wish to examine. Then to see the data you choose the Browse Data tab. This is shown in Figure 8.3.

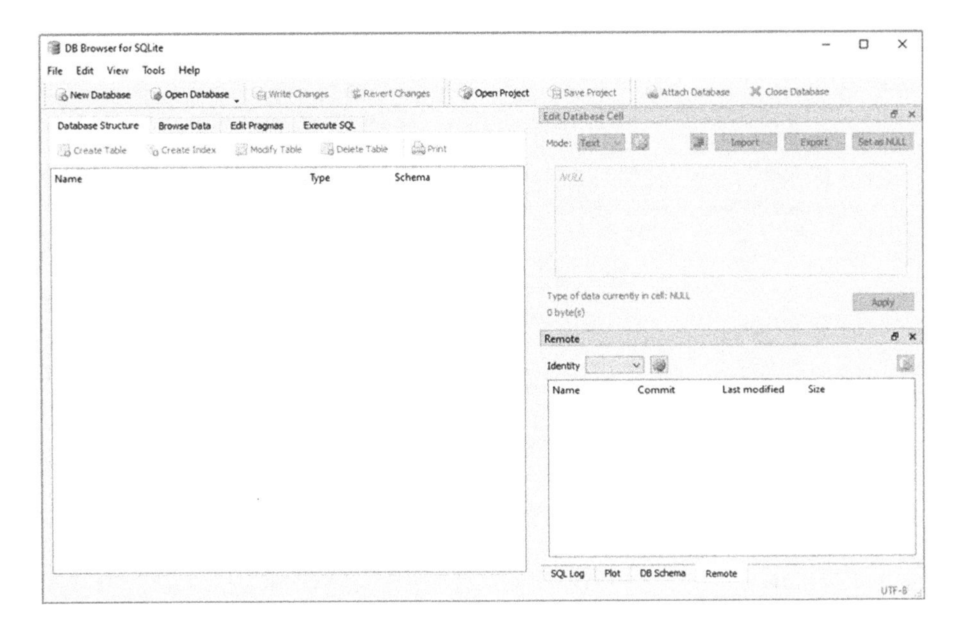

FIGURE 8.2 SQLite browser.

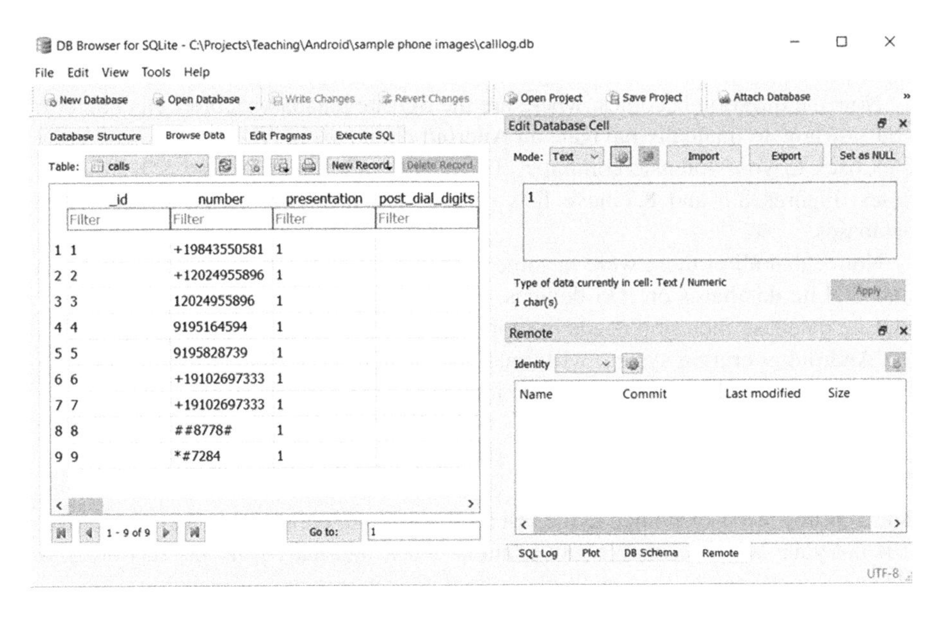

FIGURE 8.3 SQLite viewing data.

The Execute SQL tab allows you to enter SQL queries and retrieve the data you are trying to find. This is shown in Figure 8.4.

It frequently occurs that you are trying to browse data, but do not see any data. That is most likely because you have not selected the right table. There is a

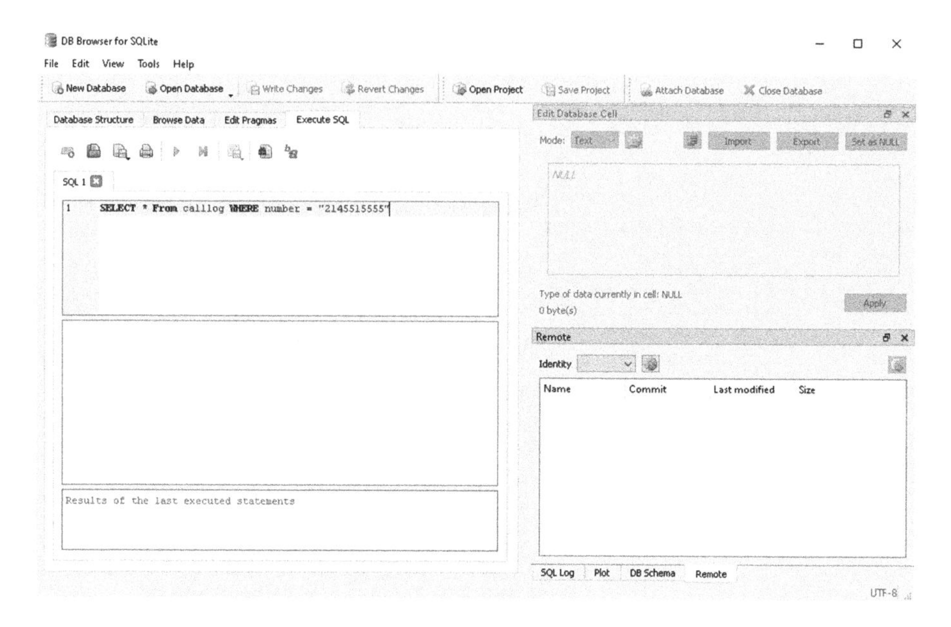

FIGURE 8.4 SQLite execute SQL.

drop-down box on the Browse Data table that allows you to select which table you wish to view. That is shown in Figure 8.5.

Now the question becomes, where are the SQLite databases stored. In chapter 6 you saw how to manually navigate an Android phone using ADB, and how to copy files back to your forensic computer. This should allow you to find SQLite databases. Figures 8.6 and 8.7 have lists of common locations for Android SQLite databases.

Notice that all of these were in some subfolder of /data/. Apple makes it easy to find SQLite databases on iOS devices. Programmers are encouraged to store any SQLite database their app needs in (application_home)/Library. Either the iOS or the Android operating system will make use of SQLite databases for storing data for apps. This makes these databases interesting forensic artifacts.

OTHER VIEWERS

The SQLite viewer examined in the last section is quite popular and easy to use. But it is not your only option. We will examine a few alternatives in this section. It is always a good idea to have multiple options for any process or technique.

SQLite Viewer for Chrome

There is an extension for Chrome that allows you to view SQLite databases. You can find it at https://chrome.google.com/webstore/detail/sqlite-viewer/golagekponhmgfoofmlepfobdmhpajia/related?hl=en. Once you have added the extension to your Chrome browser, it has a very easy to use interface, shown in Figure 8.8.

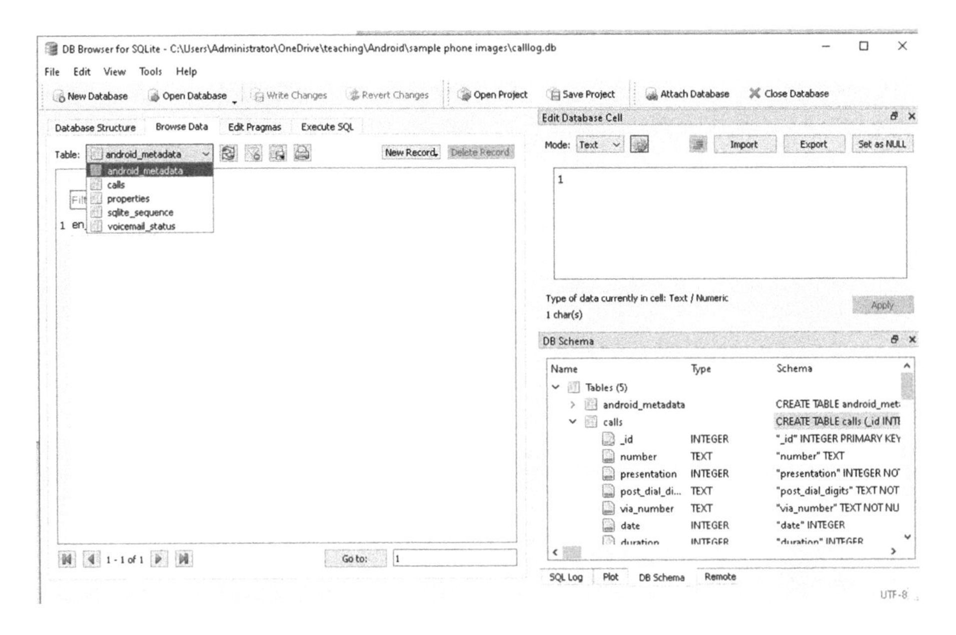

FIGURE 8.5 SQLite select table.

Call logs register	/data/data/com.android.providers.contacts/databases/contacts2.db
Call logs register (Samsung)	/data/data/com.sec.android.provider.logsprovider/databases/logs.db
Default Browser Passwords	/data/data/com.android.browser/databases/webview.db
Default Browsers History	/data/data/com.android.browser/databases/browser2.db
Dolphin Web Browser History	/data/data/mobi.mgeek.TunnyBrowser/db/browser.db
Facebook App messages	/data/data/com.facebook.katana/databases/threads_db2
Facebook Messenger messages	/data/data/com.facebook.orca/databases/threads_db2
Google Chrome History	/data/data/com.android.chrome/app_chrome/Default/History
Google Chrome Login Data (Passwords)	/data/data/com.android.chrome/app_chrome/Default/Login Data

FIGURE 8.6 SQLite database locations Part 1.

When you open an SQLite database you will be prompted to select what table you wish to view. You will also be given a list of the available tables in that database. This is shown in Figure 8.9.

You will then see the table data as shown in Figure 8.10.

You can execute SQL queries with this tool You may note that bottom left-hand corner has an SQL query already showing. One that simply selects all data from the table. You can edit that query to make your own selection. This is shown in Figure 8.11.

Kik Messenger chat messages	/data/data/kik.android/databases/kikDatabase.db
MeowChat Messages	/data/data/com.minus.android/databases/com.minus.android
Phonebook Contacts	/data/data/com.android.providers.contacts/databases/contacts2.db
Samsung SMS snippets	/data/data/com.sec.android.provider.logsprovider/databases/logs.db
Skype Calls / Messages	/data/data/com.skype.raider/files/<account_name>/main.db
SMS messages	/data/data/com.android.providers.telephony/databases/mmssms.db
Synchronised Accounts	/data/system/users/0/accounts.db
Tinder messages & users	/data/data/com.tinder/databases/tinder.db
Viber calls register	/data/data/com.viber.voip/databases/viber_data
Viber chat messages	/data/data/com.viber.voip/databases/viber_messages
WhatsApp Contacts	/data/data/com.whatsapp/databases/wa.db
WhatsApp Messages & Calls	/data/data/com.whatsapp/databases/msgstore.db
Wi-Fi passwords (WPA-PSK/WEP)	/data/misc/wifi/wpa_supplicant.conf

FIGURE 8.7 SQLite database locations Part 2.

FIGURE 8.8 SQLite viewer.

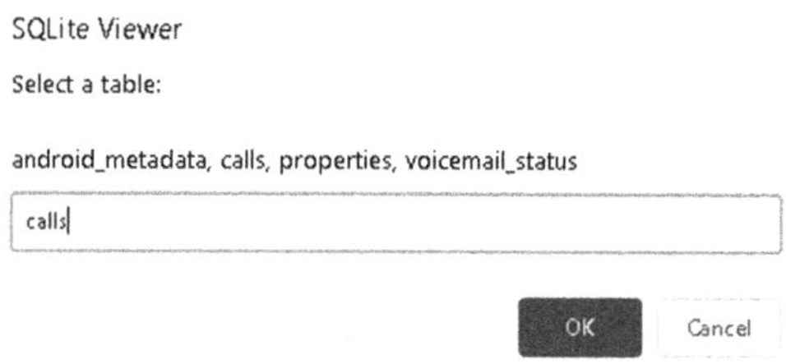

FIGURE 8.9 SQLite viewer select table.

calllog.db

_id	number	presentation	post_dial_digits	via_number	date	duration	data_usage	type	features	subscription_component_name
1	+19843550581	1			1550876757696	0		3	0	com.android.phone/com.android.services.telephony.TelephonyConnectionService
2	+12024955896	1			1552677291490	0		3	0	com.android.phone/com.android.services.telephony.TelephonyConnectionService
3	12024955896	1			1552677360000	32		4	0	com.android.phone/com.android.services.telephony.TelephonyConnectionService
4	9195164594	1			1553789026721	0		3	0	com.android.phone/com.androi calllog.db lephony.TelephonyConnectionService
5	9195828739	1			1554478621088	104		2	0	com.android.phone/com.android.services.telephony.TelephonyConnectionService
6	+19102697333	1			1554479919012	118		1	1	com.google.android.apps.tachyon/com.google.android.apps.tachyon.telecom.TachyonTel
7	+19102697333	1			1554480078699	78		2	0	com.google.android.apps.tachyon/com.google.android.apps.tachyon.telecom.TachyonTel
8	##8778#	1			1554483676788	0		2	0	com.android.phone/com.android.services.telephony.TelephonyConnectionService
9	*#7284	1			1554483692238	0		2	0	com.android.phone/com.android.services.telephony.TelephonyConnectionService

FIGURE 8.10 SQLite view data.

SQLite Viewer - Chro × | Extensions - SQLite V × | SQLite Viewer × | SQLite Viewer ×

SQLite Viewer | chrome-extension://golagekponhmgfoofmlepfobdmhpajia/data/viewer/index.html

calllog.db

_id	number	presentation	post_dial_digits	via_number	date	duration	data_usage	type	features	subscription_co
4	9195164594	1			1553789026721	0		3	0	com.android.phone/com.android.services

```
select * from calls WHERE number = 9195164594
```

FIGURE 8.11 SQLite SQL query.

SQLite Forensic Explorer

This tool is available from https://www.sqliteviewer.org/database/. There is a free version and full version. The full version is $149. You can see the main screen for this tool in Figure 8.12.

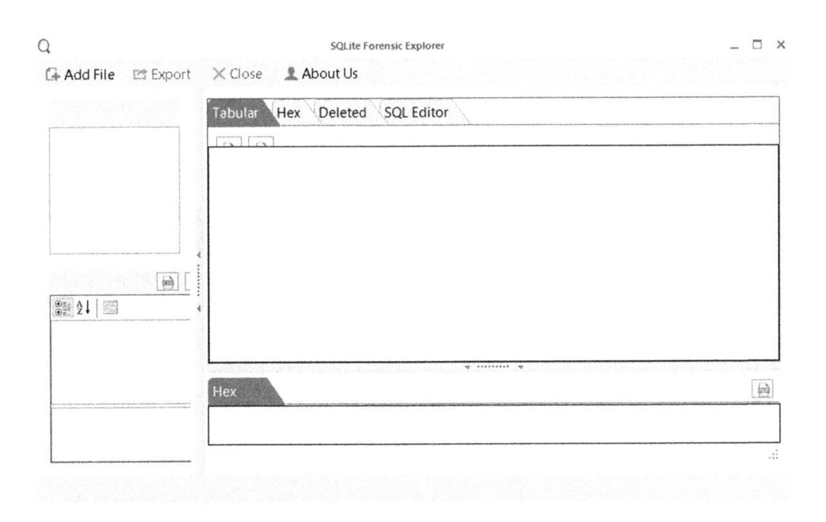

FIGURE 8.12 SQLite forensic explorer.

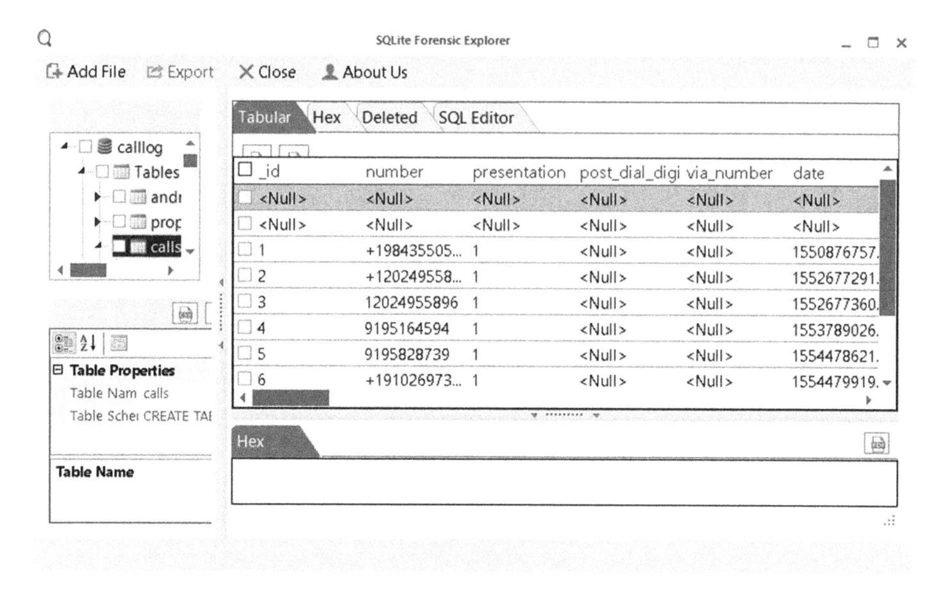

FIGURE 8.13 SQLite forensic explorer table view.

To view a particular SQLite database, you select the Add File button at the top of the screen then navigate to the location of the file. The tables will show up on the left-hand side. You then select the table you wish to view, and its contents will show in the main area of the screen. This is shown in Figure 8.13.

These are three different options for viewing SQLite databases. Whatever tool you choose, it should be clear that SQLite databases are an important part of mobile device forensics. When your phone forensic tools cannot gather data from a given app, you can always try to extract data directly from that app's SQLite database.

SQLite Viewer

This is another online SQLite database viewer. The website is https://inloop.github.io/sqlite-viewer/. It is remarkably easy to use. You drag an SQLite database to the website and then select the table you wish to view. The initial screen is shown in Figure 8.14.

Once you select a table, you will see the contents of that table as shown in Figure 8.15.

You can then execute SQL statements to refine what you are reviewing.

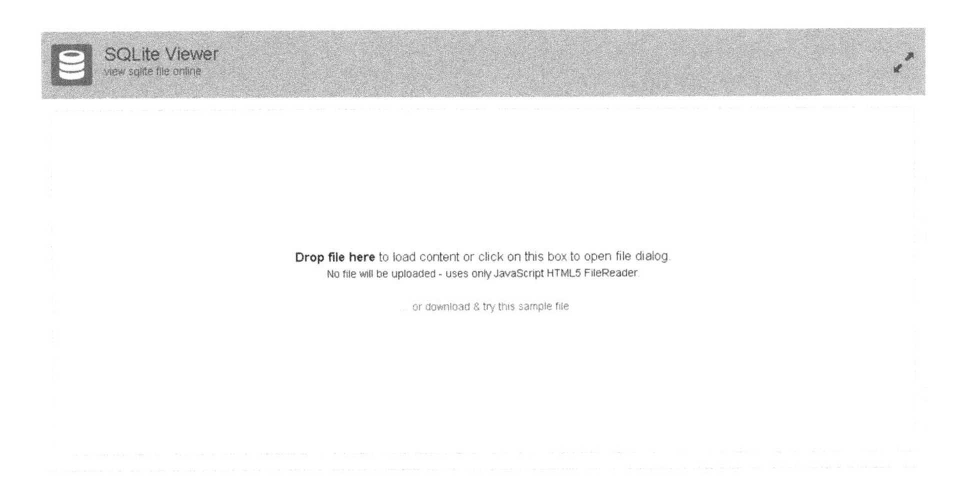

FIGURE 8.14 SQLite viewer (online).

SQLite Viewer
view sqlite file online

Drop file here to load content or click on this box to open file dialog.

calls

```
SELECT * FROM 'calls' LIMIT 0,30
```

Execute

_id	number	presentation	post_dial_digits	via_number	date	duration	data_usage	type	features	subscription_component...	subscrip
1	+19843550581	1			1550876757696	0	null	3	0	com.android.phone/com.an...	8914800(
2	+12024955896	1			1552677291490	0	null	3	0	com.android.phone/com.an...	8914800(
3	12024955896	1			1552677360000	32	null	4	0	com.android.phone/com.an...	8914800(
4	9195164594	1			1553789026721	0	null	3	0	com.android.phone/com.an...	8914800(
5	9195828739	1			1554478621088	104	null	2	0	com.android.phone/com.an...	8914800(
6	+19102697333	1			1554479919012	118	null	1	1	com.google.android.apps.t...	0
7	+19102697333	1			1554480078699	78	null	2	0	com.google.android.apps.t...	0
8	##8778#	1			1554483676788	0	null	2	0	com.android.phone/com.an...	8914800(
9	*#7284	1			1554483692238	0	null	2	0	com.android.phone/com.an...	8914800(
10	null	1			1554483692238	null	null	1	0	null	null
11	null	1			1554483692238	null	null	1	0	null	null

FIGURE 8.15 SQLite viewer (online) data view.

SYSTOOLS SQLITE VIEWER

The company Systools overs a free and a professional version of their SQLite viewer. These products can be found at https://www.systoolsgroup.com/sqlite-viewer.html. The interface looks a great deal like SQLite Forensic Explorer. You can see this in Figure 8.16.

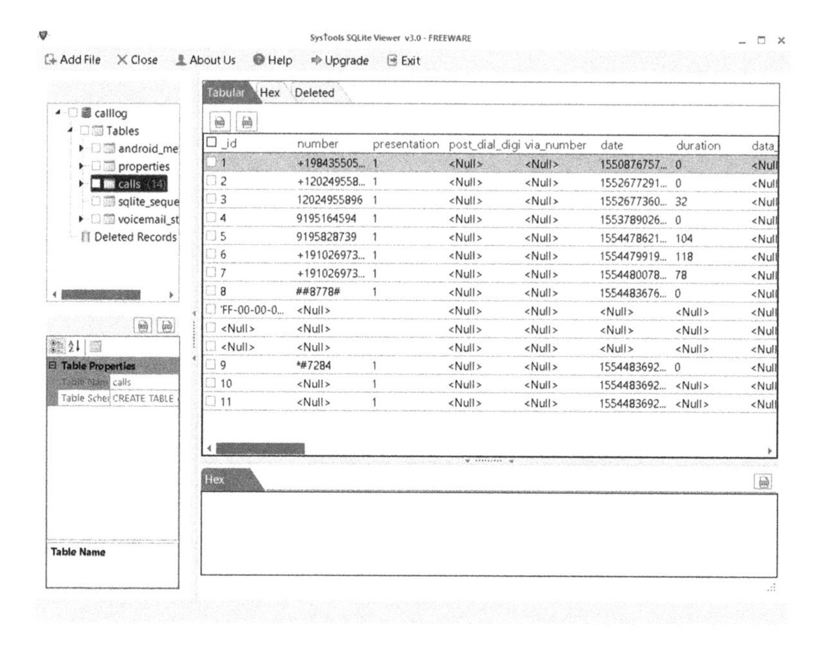

FIGURE 8.16 Systools SQLite viewer.

CHAPTER SUMMARY

Many apps use SQLite databases to store data. That makes these databases an interesting place to look for evidence. No forensic tool can get data from every one of the thousands of apps available. This may necessitate you manually locating the app in questions SQLite database, copying it to your forensic workstation, and examining it yourself. This can often find evidence you would not otherwise have. You should become comfortable with at least one or two of the SQLite viewers described in this chapter, and with basic SQL statements.

CHAPTER 8 ASSESSMENT

1. Records in a relational database are identified by ______
   a. Record number
   b. Foreign key
   c. Record key
   d. Primary key
2. The statement SELECT * FROM Employees WHERE City='Plano' and LastName!='Smith' does what?
   a. Retrieves all records in the city of Plano with a last name of Smith.
   b. Retrieves all records in the city of Plano or with a last name of Smith.
   c. Retrieves all records in the city of Plano and not the last name of Smith.
   d. Retrieves all records that are not in the city of Plano or not the last name of smith.

3. By default, the ORDER BY clause sorts in what order?
   a. Ascending
   b. Descending
   c. There is no default you must set it.
   d. It varies.
4. SQL Lite stores decimal numbers as what type?
   a. Float
   b. Double
   c. Real
   d. Decimal

CHAPTER 8 LABS

Lab 8.1

Download an SQLite database sample or retrieve one from a phone. You can get samples from

https://www.sqlitetutorial.net/sqlite-sample-database/

https://github.com/jpwhite3/northwind-SQLite3

Whatever database you use, choose one of the tools described in this chapter. Load the database in that tool and view the tables. Execute at least three SQL Queries on the database. The Point of this lab is to get comfortable with at least one of the tools and with basic SQL Queries. You should consider using all three tools but use at least one.

9 Cell Site Analysis and Smart TV Forensics

In addition to examining the actual mobile device, one may need to analyze cell phone records from the cellullar company. This process is called cell site analysis. It is important for the forensic examiner, to understand what cell site analysis can and cannot do. It is important to understand the limitations and be able to work within those limitations.

INTRODUCTION

This chapter combines two topics that are peripheral to mobile device forensics, but related. The first is cell site analysis. It is common for forensic examiners to receive call history records. These can be analyzed to retrieve some information. However, the information available may not be as detailed as some analysts believe. A second issue addressed in this chapter is forensic examination of smart televisions. Many smart TVs use Android operating system, or some close variation. That makes smart TV analysis relevant to Android forensics.

CELL SITE ANALYSIS

Sometimes mobile device forensics is augmented by examining cell usage records from the carrier. And in some cases, those cell usage records are all that is available. This is true if the phone has been lost or destroyed. Cell phone records acquired from cell phone carriers will reveal the particular tower that an individual cell phone is connected to at a specific time. This process is usually referred to as Cell Site Analysis. These records have been used to attempt to locate individuals in both criminal and civil investigations. Some carrier records will include data upload and download.

While this information can be vital in any investigation, it is important for the analyst to proceed with caution. Too often forensic examiners attempt to use carrier records to support conclusions that simply are beyond what the records can support. Consider location information. The cell carrier records can tell you what cell tower you were connected to, but probably cannot get any more specific on location than that. Depending on the area you are in, and the density of towers in that area, this may or may not be particularly helpful.

Cell towers have a general range of 5 to 10 kilometers. However, in some cases the range can extend for 10's of kilometers. However, the tower's range can be diminished by factors in the terrain including either extensive mountains, particularly those with metal ores, or large urban constructions. It is also true that when there is a high density of cell phone towers in a given geographical region, that a cell phone will be handed off to another tower when in range of that second tower. Thus, cell tower density can reduce the range that a given cell tower covers. Generally speaking, the smaller the cell tower's range, the more meaningful is locating an individual in the range of that cell tower.

DOI: 10.1201/9781003118718-9

Consider two examples of cell tower density that are on opposite ends of that density spectrum. Cell tower density in Manhattan compared to cell tower density in Montana. Figure 9.1 shows cell tower locations in Manhattan.

It is quite obvious that this is a rather dense concentration of cell phone towers. In many cases, there are towers within a few city blocks of each other. This means that identifying that a given mobile device was connected to a particular tower, at a particular time, much more narrowly identifies the location of the cell phone. In one extreme case, there are two cell phone towers within the same city block. This is shown in Figure 9.2.

FIGURE 9.1 Cell phone towers in Manhattan.

FIGURE 9.2 Two cell towers in one block.

FIGURE 9.3 Cell towers in Basin, Montana.

Obviously, if cell site analysis shows a person connected to one of these towers at a specific time, that is a rather narrow geographical area. This can be very useful information in a forensic investigation. However, there are other areas where cell phone density is quite sparse. As one example, Basin, Montana, is shown in Figure 9.3.

Deterring that a given mobile device was connected to this cell tower at a particular time does very little to determine the user's location with any precision. This lack of precision in cell tower records is one reason such records cannot be used to definitively identify a person's location. There have been cases of forensic examiners attempting to use these records to state a suspect was at the scene of a crime at the time the crime was committed. However, the lack of accuracy in these records makes such conclusions untenable.

In general, a location defined by latitude and longitude, found in historical cell phone records, can only identify that the cell phone was within range of a specific cell phone tower. Given that a cell tower can have a range between 5 kilometers to 10 kilometers, and occasionally beyond that provides a large area in which the device could be. Such records can be useful in determining that a suspect was not in the vicinity of a crime. But they cannot definitely place the suspect at the crime scene. Even in the rather dense area of Manhattan, at most the cell site analysis can tell you that the suspect was on the same block as the crime being committed. This can still be useful, however, if the suspect has made statements that they were not in the area.

While cell tower ranges tend to be in the 5 to 10 km range, the National Institute of Standards (NIST) states that cell phone towers can be servicing phones as far away as 35 kilometers (21.74 miles). This presents a very large area where a mobile device could be in the area and be connected to that tower. NIST is very clear in what cell phone usage records can, and cannot demonstrate:

TABLE 9.1
Cell Phone Tower Area Covered

Radius (km)	Radius (miles)	Area (km)	Area (miles)
5	3.16	78.5	31.35
10	6.32	314	121.2
35	21.74	3846.5	1484.05

> While plotting call record locations and information onto a map can sometimes be useful, it does not necessarily provide a complete and accurate picture. Cell towers can service phones at distances of up to 35 kilometers (approximately 21 miles) and may service several distinct sectors.[1]

The NIST standards document goes on to state that these records can only give you a general area a phone was in, and a general direction of motion. The preceding quote should clarify the problematic nature of attempting to geolocate a particular phone based solely on historical cell phone records.

To better understand the issue with how much area is covered with a single cell tower, it is helpful to calculate how much a given radius covers. Utilizing the formula for area of a circle, one can plot how much area is covered by a given cell phone tower, based on estimates of coverage radius. This is shown in Table 9.1.

$$A = \pi r^2$$

It should be noted that Table 9.1 uses a variety of different radii for cell towers, because varied sources provide a range of estimates of the appropriate radius. But consider even the smaller radius of 5 km/3.16 miles. That covers an area of over 31 square miles. Identifying a suspect as being within a 31 square mile area really does not do much to tie him or her to a given crime. It can, however, provide an alibi.

Even when using live GPS, the Federal Communications Commission (FCC) states "GPS accuracy varies and could incorrectly place the victim's location at their neighbor's home." and the document continues with "Using cell towers to detect location is not as accurate as GPS. Locating a mobile phone based on a single cell tower can place the mobile phone in a broad area, but it cannot pinpoint it."[2]

The data currently available via numerous studies, clearly demonstrates that historical cell phone data, cannot be used to pinpoint the precise location of a given cell phone. It can only provide the general area in which the phone was located at a particular time.

Various researchers have published studies bringing into question the accuracy of location information derived from cell site analysis. Thomas O'Malley discusses the fact that historical call records can only be used to place a given phone within a general vicinity.[3] Coutts and Shelby are even more detailed, stating "Recourse to the CCR Cell ID is useful only to establish a possible area in which a sending mobile was locate when a call was made from that mobile and 'picked up' by a receiving telephone."[4] Herbert Dixon goes even further in his article "Scientific Fact or Junk Science; Tracking a Cell Phone without GPS" describing the use of cell phone usage records to determine location of a cell phone as "junk science".[5]

What is clear from both reviewing existing studies and looking at the actual range of cell towers is that historical cell call data are not accurate and do not lead to scientifically valid conclusions. Such historical records, however, can be useful in impeaching a suspect. If a suspect claims that he or she was in a different geographic area when the incident occurred, historical cell phone records can refute that claim. It is also possible to use cell site analysis to get a general pattern of movement and average speed.

Effective Use of Cell Tower Records

A process for deriving general data from cell site analysis was first published in 2018 and is given here in this section.[6] The first step in this process is to plot cell phone tower connection against time. However, this must be done understanding the range limitations, previously discussed in this chapter. Each latitude and longitude in a given historical cell phone record represents a tower and a radius around that tower.

If detailed data regarding cell tower signal strength and signal obstacles is available, then it is possible to calculate the point that a cell phone is handed off from one tower to the next. However, this level of detail may not always be present in cell tower records. Therefore, a method of estimation is used. The assumption is that the cell phone will be passed from one tower to the next at the midway point between towers.

While there are different types of cell phone handovers, when contemplating historical cell phone records, the handover between towers is the one to focus on. The MAP protocol (Mobile Application Part) is used to move from one MSC (Mobile Switching Center) to another. As a general estimate, the midway point between the two cell towers is an effective hypothetical handoff point.

Cell phone records show a latitude and longitude when a specific call is made. Therefore, if the handoff occurs during a call, both towers' latitude and longitude will be recorded. For this current methodology to work properly there must be at least three points (i.e., cell tower latitude and longitude values) to plot. In the following demonstration of the method assumes the data presented in Table 9.2.

The call begins while the cell phone is within range of tower A, and continues through tower B and tower C. Using the data in table II , it can be seen that transitioning from tower A to tower B should occur at .75 km from tower A. It is also apparent that this occurred at the 1 minute and 02 seconds (1.033 minutes) after the call begins. This provides a mean velocity of .726 km/s. However, there is a margin of error. The call could have been initiated just barely within the opposite side of cell tower A's radius, or just prior to the handoff to cell tower B. This means the distance traveled could vary from as little as .1 km/1.03 minutes up to 1.48 km/ 1.03 minutes. The reason the second number is 1.48 km is that is assuming the greatest possible distance, from just .1 km within the far side of cell tower A radius,

TABLE 9.2
Representative Cell Tower Data

Time	Cell Tower Connected To	Distance to Next Tower in The Path	Mean Distance to Handoff to Next Tower
01/01/2021 09:12:52	A	1.5 km	.75 km
01/01/2021 09:14:54	B	2.0 km	1.0 km
01/01/2021 09:17:52	C	2.0 km	1.0 km

to just before the hand off to cell tower B. This represents a rather large margin of error. This is why additional points along the path will be more useful in narrowing the margin of error. It must also be noted that the first cell phone tower will have the greatest margin of error in location. This is due to the fact that the call could have been initiated from anywhere within the cell tower radius. When a call is in progress, and passes through three, or more, cell phone towers, it is clear for each intermediate tower, when the phone entered that tower's radius and when it left that cell tower's radius and was handed off to the next tower.

However, even with just the first tower, the range of possible velocities can be reduced, thus reducing the margin of error. Beginning with the shortest possible distance of .1 km in 1.03 seconds, this is equivalent to 5.82 km/hour. This is a reasonable speed for a person walking fast. The other extreme of 1.48 km in 1.03 minutes is equivalent to 86.21 km/hour. Clearly this is impossible for a person not in a vehicle. So, the next step would be to consider the roads that could be traversed in the geographical area in question, as well as average speed of traffic on those roads.

Continuing with the representative data in Table 9.2, the next tower (tower B) has a negligible margin of error. It is known that the cell phone entered its radius and was handed off from tower A to tower B at 9:14:54, then was handed off to tower C at 9:17:52, 2 minutes and 58 seconds later. From entering the radius of tower B to the center of that radius is .75 km. Then it is an additional 1 km to hand off to tower C. This provides a distance of 1.75 km covered in 2 minutes and 58 seconds (2.96 seconds). Thus, the average velocity was .678 km/second or 40.69 km/hour. However, the margin of error must also be calculated. The cell towers after the first one have minimal margins of error. This is because the call is active and is passing from one cell tower to the next. The first call has the largest margin of error, because the call could have been initiated anywhere within the radius of the call phone tower.

This provides a mean velocity for the cell phone that has a reasonable margin of error. This average speed can be compared to the wide range of average speeds using just cell tower A, and it is more likely that the average speed in cell tower A's radius was at least similar to that in cell tower B's radius. This data would of course be subject to analysis of traffic patterns and roads.

As can be seen with three towers, the mean velocity of the phone can be calculated with much greater accuracy than just a single tower. If additional towers are available during the span of a single call, then this accuracy can be improved even more. Alternatively, additional call data within a short period of time of this call, can also be analyzed in order to further refine the average speed, with an ever-decreasing margin of error.

Put formally, the calculation would involve summing the distance from hand off to hand off along with calculating the aggregate margin of error. This is shown in Equation 9.1.

$$\mathrm{E} = \sum_{i=1}^{n} D_i +/- \sum_{i=1}^{n} M_i \tag{9.1}$$

In Equation 9.1, D_i is the mean distance for hand-off and E is the margin of error.

Thus far the calculations have assumed a straight line between points. This does provide a general estimate for distances and speeds. Coupled with multiple towers, this methodology can set upper and lower bounds for travel speed, as well as location.

In order to further refine the calculations, and to reduce the margin of error, the next step is to attempt to map this to the most likely travel routes. The estimate of speed calculated previously is useful in this phase. If the speed is faster than a human can reasonably travel by foot, then one can assume a vehicle was used. This can reduce the possible paths based on roads available or trains. For example, consider the two cell towers in Manhattan, shown in Figure 9.4.

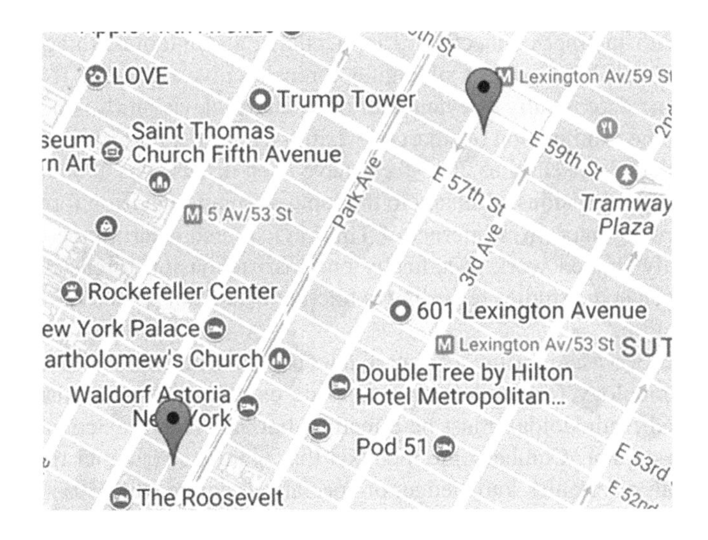

FIGURE 9.4 Exemplary cell phone towers.

Assuming the estimates of speed indicate travel via vehicle, there are two most likely paths:

47th street to Park avenue to 58th street, to the final destination
47th street to Lexington avenue to the final destination.

Other, more circuitous paths would have either required more time to traverse, or increased speed. If the increased speed or time required for other routes is beyond the range of parameters already calculated, then those routes can be ignored.

This methodology is improved not only with additional call data, but also with additional information regarding roads, traffic conditions, and other relevant data that might narrow the possible geographic limits involved in calculating both location and velocity.

The issue is actually quite simple. Cell site analysis does not reveal the level of geographic detail that some analysts have tried to derive from it. However, it can be quite useful for determining general location of a suspect, eliminating a suspect, or calculating general travel patterns. This type of evidence is unlikely to be definitive in any given case but can be important ancillary evidence.

SMART TELEVISIONS

This may seem like an odd section in a book on mobile devices. However, there are two reasons this is included here. The first is that many smart TVs use Android as their operating system. Those that don't, typically have some other Linux variant. A second reason is that this topic has come up frequently in training sessions I have conducted on mobile forensics. Many forensic examiners are now finding smart TVs to be part of the available evidence.

Smart televisions have become ubiquitous. It is important to bear in mind that these devices are not merely televisions, but rather they are fully functioning computers with internet connectivity, applications, and internal storage. As such, these devices can be a subject of digital forensics investigations. It is certainly possible to find forensically relevant data on a smart television. It is also that case that ignoring the examination of smart televisions can lead a forensic analyst to miss valuable evidence. As early as 2013 there have been discussions about forensically analyzing smart televisions.[7] Related to the topic of smart television forensics, there has been growing interest in Internet of Things (IoT) and smart home forensics.[8],[9] However, only limited work specifically on smart television forensics. However, there have been some studies regarding extracting data from related devices such as the Amazon Fire.[10]

Given the pervasive nature of smart televisions, what is needed is a clearly defined methodology for conducting a forensic examination of a smart TV. Any such proposed methodology must be effective, but also must be actionable for the forensic investigator. Cumbersome methods that require substantial time commitment, in-depth electronics knowledge, or specialized equipment will not be usable for many forensic investigators. Therefore, a methodology that is forensically sound and easily executed is required. The current paper posits such a methodology.

Methodology

As of 2020, the majority of smart televisions utilize the Android operating system, or some variation thereof. Android TV is used by Toshiba, Asus, LG, Sony, Sharp, and many other vendors. Google TV was previously utilized by several vendors, including LG and Sony, but was supplanted by Android TV. Roku TV uses a custom Linux distribution named Roku OS. Given that Android is based on Linux, many techniques used for Android TV will also work on Roku televisions. Some LG televisions utilize webOS, but it is also a Linux-based operating system. Samsung has utilized Tizen and Oray as operating systems. However, both of those are Linux-based systems.

The data regarding widely used smart television operating systems strongly suggests that focusing on Android and variations will provide a broad-based methodology that can be utilized with a wide array of televisions. The bulk of that methodology will also be applicable to other Linux-based television operating systems. Utilizing an Android-based approach will provide a methodology that is applicable to the largest number of different brands of smart TV.

There are two approaches that meet the requirements of being forensically sound and are readily usable by a broad range of forensic examiners. The first is using Android Debugging Bridge (ADB). ADB is already widely accepted for forensics. Therefore, utilizing ADB for forensic analysis is scientifically sound. Any Android TV can be accessed using ADB. Given that not all smart TVs allow a USB connection, the best way is to use ADB via network. Android Debug Bridge (adb) is a versatile command-line tool that lets you communicate with a device. The adb command facilitates a variety of device actions, such as installing and debugging apps, and it provides access to a Unix shell that you can use to run a variety of commands on a device.

Using ADB

The first step is to install the App Developer Tools. This is a free app, that is already included on some televisions. It provides the examiner the same options one would find on an Android phone. This is needed to turn on USB debugging. This is shown in Figure 9.5.

After USB debugging is turned on, the next step is to connect to the television from a forensic workstation. The workstation only needs the Android Debugging Bridge installed, which is a free download (e.g., https://developer.android.com/studio/command-line/adb).

Connecting is a straightforward process. As shown in Figure 9.6, from the command line, one uses the following command (replacing the IP address with the IP address of the smart TV in question):

adb connect 192.168.1.85:5555.

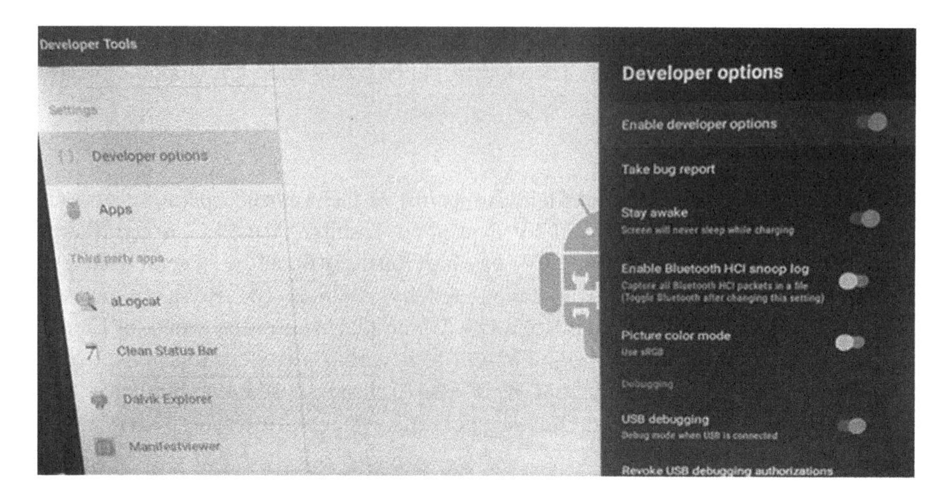

FIGURE 9.5 Smart TV USB debugging.

```
:\platform-tools>adb connect 192.168.1.85:5555
lready connected to 192.168.1.85:5555

:\platform-tools>adb devices
ist of devices attached
92.168.1.85:5555        device

:\platform-tools>adb shell
anda:/ $ ls
cct        charger default.prop init                      init.usb.configfs.rc mnt  product sys
in         config  dev          init.environ.rc           init.usb.rc          odm  sbin    system
ugreports  d       etc          init.rc                   init.zygote32.rc     oem  sdcard  ueventd.rc
ache       data    factory      init.recovery.m5621.rc    lost+found           proc storage vendor
anda:/ $
```

FIGURE 9.6 ADB connect via IP.

Connecting via IP is used because many smart TVs limit the USB port, and you may not be able to connect via USB. By typing *adb devices* one can verify there is a connection. Then executing *adb shell*, provides a shell on the target device and the ability to execute a range of Linux commands. The shell can be exited at any time by typing *exit*. A first step, done from without the shell, is to back up the device:

adb backup -all -f C: \backup.ab.

After backing up the target device, it is necessary to document details about the device. The *adb shell getprop* command can be useful for this purpose, as shown in Figure 9.7.

There are other commands that should be utilized in an investigation. Some commands you may find useful are shown in Table 9.3.

```
D:\platform-tools>getprop ro.product.model
'getprop' is not recognized as an internal or external command,
operable program or batch file.

D:\platform-tools>adb shell getprop ro.product.model
2K Smart TV

D:\platform-tools>adb shell getprop ro.build.version.release
9

D:\platform-tools>adb shell getprop ro.serialno
AC86897C49
```

FIGURE 9.7 Get TV data.

Many of these commands list a prodigious amount of information. It can be useful to export the data to a text file, such as by using:

adb shell dumpsys > data.txt

By creating a backup, and gathering data via the android debugging bridge, the examiner can fully examine the smart TV. Using *adb pull*, any interesting data on the television can be pulled. For example, photos, videos, or other files. The list provided here are the essential commands that will be utilized to perform a standard examination of the smart TV. However, if the examiner is well versed in ADB commands, there are additional commands that might be useful.

In some cases, the examiner may have difficulty connecting to the target device. In these cases, it often is useful to end the adb server on the forensic machine and restart it. This is accomplished by first running the kill-server command, then adb devices, as shown in Figure 9.8.

This causes the adb service to end, but then restarts it. Particularly if you have attached to numerous devices, or if you have other software on your computer that uses adb, this can be necessary. Many phone forensics software packages actually use adb to perform their forensic exam.

This general overview of how to use Android Debugging Bridge to obtain forensically relevant information from a Smart TV, leads naturally to a need for a step-by-step process. This is a process that can be generally applied to any Android TV. The steps are as follows:

Step 1: Turn on USB Debugging on the target device.
Step 2: Connect to the smart TV from a forensic workstation using *adb connect 192.168.1.85:5555.*
Step 3: *Backup the device using adb backup -all -f C: \backup.ab*
Step 4: Utilize *adb shell getprop* to gather details about the smart tv. These details are necessary for thoroughly documenting the forensic investigation. It may be useful to get all properties via *adb shell getprop > properties.txt.*

TABLE 9.3
ADB Commands for Smart TV Forensics

Command	Purpose
adb pull	Pulls a single file or entire directory from the device to the connected computer. For example: adb pull /sdcard/screen.png
adb shell dumpsys	This is a very versatile command with several options:*adb shell dumpsys package com.android.chrome* will dump all the data for a given package. *adb shell dumpsys* activity provides information about Activity Manager, activities, providers, services, broadcasts, etc.
adb shell netstat	List network connections. This can indicate the presence of a remote connection to the TV, including spyware. This is the standard netstat command widely used on desktop operating systems.
adb shell list packages -f	See packages and the associated files. This will provide a list of everything installed on the Smart TV.
adb shell getprop	One can gather a greate deal of information about the smart TV with commands such as: adb shell getprop ro.product.model adb shell getprop ro.build.version.release adb shell getprop ro.serialno adb shell getprop ro.board.platform adb shell getprop ro.build.version.security_patch
shell commands	Several shell commands will provide information about the target device, such as: adb shell pm list adb shell service list
adb logcat	This command dumps an extensive log of activity on the television. Note this is a rather long list and should be dumped to a text file.
adb shell pm list features	This command will list device features
adb shell service list	This command lists services on the device

Table 1: ADB commands

Step 5: Now utilize adb commands as needed to locate relevant data, and copy said data back to the forensic workstation using *adb pull*. The specific commands used will depend on the nature of the incident being investigated as well as the evidence found.

Following this method, along with careful documentation of all steps taken, will yield forensically sound evidence. Furthermore, this is a methodology that can be implemented entirely with free tools. Furthermore, the author of this current paper has taught digital forensics using ADB for several years. This experience has

```
D:\platform-tools>adb devices
List of devices attached
adb server version (32) doesn't match this client (40); killing...
* daemon started successfully
error: protocol fault (status 43 4e 58 4e?!)

D:\platform-tools>adb kill-server

D:\platform-tools>adb devices
List of devices attached
* daemon not running; starting now at tcp:5037
* daemon started successfully
192.168.1.85:5555       connecting
```

FIGURE 9.8 Kill server.

demonstrated that it normally takes approximately 4 hours of training to achieve proficiency for the typical forensic investigator. Consequently, this methodology is one that can be employed by almost any forensic examiner. This method, using Android Debugging Bridge, for smart television forensics, is the preferred method proposed by this paper.

AnyDesk

There can be smart televisions that are Linux based, but do not support the Android Debugging Bridge. For this reason, it is necessary to have an additional forensic modality. This alternative methodology can be implemented when ADB is not an option. There are several apps one can install on a smart TV to allow examination of the device from a remote computer. One such app is AnyDesk. This is free, though there is a paid version. When installed on a computer, one can then connect to any Smart TV on the same network that also has the AnyDesk app. The main screen for AnyDesk is shown in Figure 9.9.

To connect, you enter the remote address, this is not an IP address, but rather an AnyDesk created address. The target machine will receive a message asking whether or not to accept the connection. Once the connection is accepted, examination of the machine can begin. AnyDesk allows one to record the entire session, which is beneficial for keeping a record of forensic activity. This is shown in Figure 9.10.

AnyDesk will also allow you to navigate the file system on the target machine and copy files back to the forensic workstation. This allows the investigator to retrieve any data that is relevant to the investigation. The retrieval process operates via a standard file manager interface, as shown in Figure 9.11.

The file transfer process is the most important. As the examiner finds files relevant to his or her investigation, those files can be copied back to the forensic machine for further analysis. This overview of AnyDesk leads to a step-by-step process that can be used to forensically analyze a smart TV. The steps are as follows:

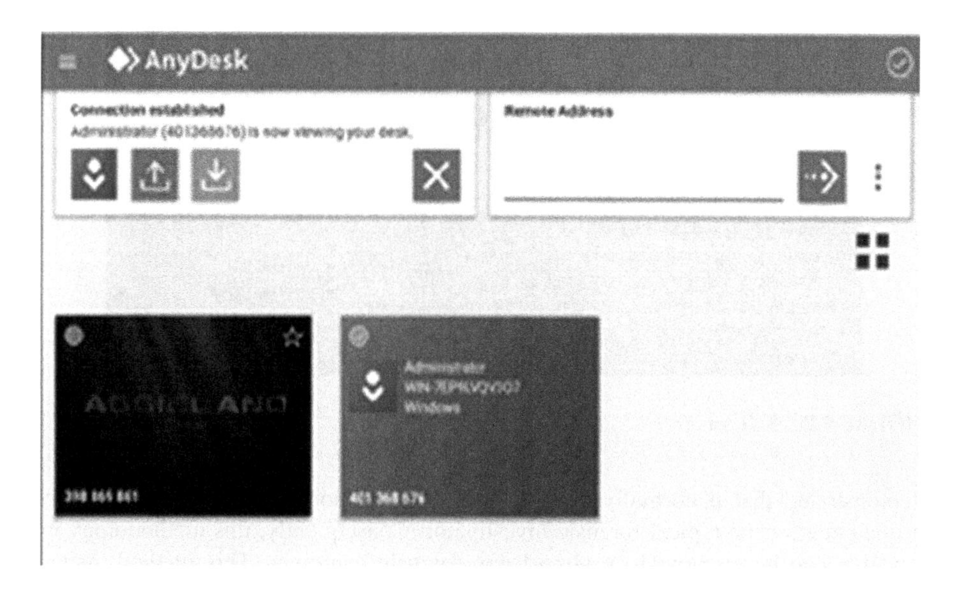

FIGURE 9.9 AnyDesk.

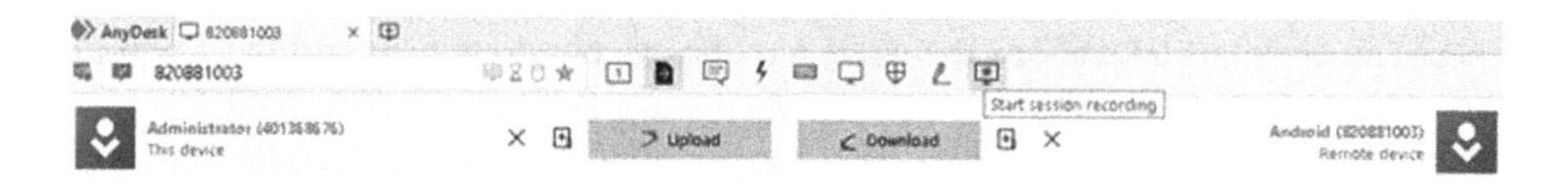

FIGURE 9.10 AnyDesk connection.

Step 1: Install AnyDesk on the target TV and on the forensic computer.
Step 2: Connect from the forensic computer to the smart TV, using AnyDesk.
Step 3: Turn on recording to record the entire session.
Step 4: Copy all relevant data to a folder on the forensic computer.

This method will provide basic forensic functionality. It is not as versatile as Android Debugging Bridge, but certainly can be a viable option when adb is not feasible. The most important step is to turn on recording so that the entire session is recorded. This is necessary to validate the integrity of the investigation.

CHAPTER SUMMARY

In chapter 9 you have seen both cell site analysis and smart TV forensics. Both of these areas of forensics are related to mobile device forensics, while not dealing directly with the mobile device. Cell site analysis uses cell phone records to attempt to locate the time and location of calls. This data can be substantially inaccurate depending on cell tower parameters. However, a methodology was presented in this chapter for making estimates, even from inaccurate data.

FIGURE 9.11 AnyDesk file manager.

Smart TV's frequently use the Android operating system. This means that many of the Android techniques you learned in chapter 6 can be used. Particularly, Android Debugging Bridge can be a useful tool. This chapter walked you through how to use ADB for Smart TV forensics, and also introduced you to using file management apps to extract data from a smart TV.

CHAPTER 9 ASSESSMENT

1. According to NIST what is the maximum range of a cell tower?
   a. 35 km
   b. 35 miles
   c. 5 to 10 km
   d. 5 to 10 miles
2. What is the FCC position on GPS data?
   a. Unlike CSA, GPS is precise.
   b. GPS is usually inaccurate.
   c. GPS is more accurate than CSA.
   d. GPS and CSA are about equally accurate.
3. The following formula is used for what purpose?
   $E = \sum_{i=1}^{n} Di + /- \sum_{i=1}^{n} M_i$
   a. Determining the margin of error for Cell Site analysis.
   b. Determining the range of a cell tower.
   c. Determining the area serviced by a cell tower.
   d. Determining position of a cell tower.
4. What operating system does Samsung use for Smart TVs?
   a. webOS and Android
   b. webOS and Tizen
   c. Android TV and Oray
   d. Tizen and Oray
5. Why use IP connection for ADB on a Smart TV?
   a. You should only use IP if you don't have a USB cable.
   b. You usually cannot connect via USB cable.
   c. It is faster.
   d. It is more reliable.

CHAPTER 9 LABS

Lab 9.1

This requires the use of a Smart TV. You can use your own, but limit what you do with the Smart TV so that there is no chance of altering the TV's configuration.

First turn on usb debugging on the TV
Then locate the TV's IP address
Connect via adb over internet.
Perform a backup of the TV.
Retrieve data such as serial number, model number, etc.

NOTES

1 https://csrc.nist.gov/publications/detail/sp/800-101/rev-1/final
2 https://www.fcc.gov/sites/default/files/911-help-sms-whitepaper0515.pdf
3 O'Malley, T. A. (2011). Using historical cell site analysis evidence in criminal trials. *United States Attorneys' Bulletin*, *59*, 16.
4 Coutts, R. P., & Selby, H. (2016). Problems with cell phone evidence tendered to prove the location of a person at a point in time. *Digital Evidence & Elec. Signature Law Review*, *13*, 76.
5 Dixon Jr, H. B. (2014). Scientific Fact or Junk Science? Tracking a Cell Phone without GPS. The *Judges' Journal*, *53*, 37.
6 Easttom, C. (2018). A Method For Using Historical GPS Phone Records. Digital Forensics Magazine, 36.
7 Al Falayleh, M. (2013). A review of Smart TV forensics: Present state & future challenges. In The International Conference on Digital Information Processing, E-Business and Cloud Computing (DIPECC) (p. 50). Society of Digital Information and Wireless Communication.
8 Li, S., Choo, K. K. R., Sun, Q., Buchanan, W. J., & Cao, J. (2019). IoT forensics: Amazon echo as a use case. *IEEE Internet of Things Journal*, *6*(4), 6487–6497.
9 Goudbeek, A., Choo, K. K. R., & Le-Khac, N. A. (2018). A forensic investigation framework for smart home environment. In 2018 17th IEEE International Conference On Trust, Security And Privacy In Computing And Communications/12th IEEE International
10 Hadgkiss, M., Morris, S., & Paget, S. (2019). Sifting through the ashes: Amazon Fire TV stick acquisition and analysis. *Digital Investigation*, *28*, 112–118.

Section III

Additional Topics

10 Anti Forensics

INTRODUCTION

It should come as no surprise that people frequently take steps to make it difficult for one to retrieve data from their phone. Many phone users take some steps purely for privacy reasons. People engaged in criminal activity may take additional steps with the goal of preventing a forensic examiner from extracting evidence from the phone. Some of these anti forensic measures can be effectively countered, others are extremely difficult to circumvent. In this chapter you will be introduced to the various methods that can be used to thwart mobile device forensics. You will also see methods that can potentially circumvent such measures.

PHONE LOCKING

Modern mobile devices offer a range of security features for the user. The most obvious is the phone lock. The unlocking mechanism can be a PIN, a pattern, facial recognition, or some phones even offer proximity unlocking. For example, Samsung offers proximity unlock. This can be configured to unlock if the phone is in proximity to a particular wireless access point, a smartwatch, or even an automobile Bluetooth signal. These features are meant for the convenience of the user. The phone unlocks at trusted locations or when near trusted devices.

There are tools that claim to be able to get around a PIN or pattern. These have debatable efficacy. Obviously, they work in some instances but not all. It is possible one of these tools might help you get into a locked phone. Some of these tools will remove the lock but will also wipe the phone and return it to factory condition. That is unacceptable for forensic purposes. A better approach is to study the suspect and determine likely pin numbers. We will briefly discuss some of these tools later in the chapter.

If the suspect is using smart lock, then taking the phone to their residence or automobile might unlock it. With facial recognition you can literally hold the phone in front of the suspect's face to unlock it. It should be noted that getting into a locked phone is a non-trivial task in most cases and you won't always be able to.

STEGANOGRAPHY

Steganography is one way to hide data on any device. There are a number of steganography software tools easily available on the internet. The ubiquitous nature of these tools means that you should expect to find steganography at some point in your forensics career.

DOI: 10.1201/9781003118718-10

FIGURE 10.1 Two dogs.

There are many techniques for accomplishing steganography, but one of the most common methods is done using the least significant bit (LSB) method. This method changes the last bit, or least significant bit, of a series of bits. Each pixel on your device is represented by 24 bits, there are three sets of 8 bits. There are 8 bits each for red, green, and blue. If you change just one bit on a pixel, the picture is not noticeably altered. Consider the picture of two dogs shown in Figure 10.1.

The two images are identical. In fact, they are literally copies of one another. Now I will use Microsoft Paint to change a pixel value. In Figure 10.2 you can see the original color selected on the left. On the right you can see the red value changed by 1.

Now after changing 1 bit for a given color in one of the two dog pictures, view the dogs again, this is shown in Figure 10.3.

You cannot tell the difference. Even if you enlarge the image, zoom in, view it any way you wish. A change of 1 bit, even in many pixels, is not detectable by the human eye. This is the foundation for modern image steganography. If you change the least significant bit in a pixel, the image still looks the same. However, pictures are made up of thousands, and in some cases millions, of pixels. Changing the least significant bit allows you to hide thousands or millions of bits of data.

With that basic conceptual understanding of steganography now firmly in your mind, we can turn to some general terminology. Regardless of the steganographic tool you use, there are some terms that are common to the field of steganography:

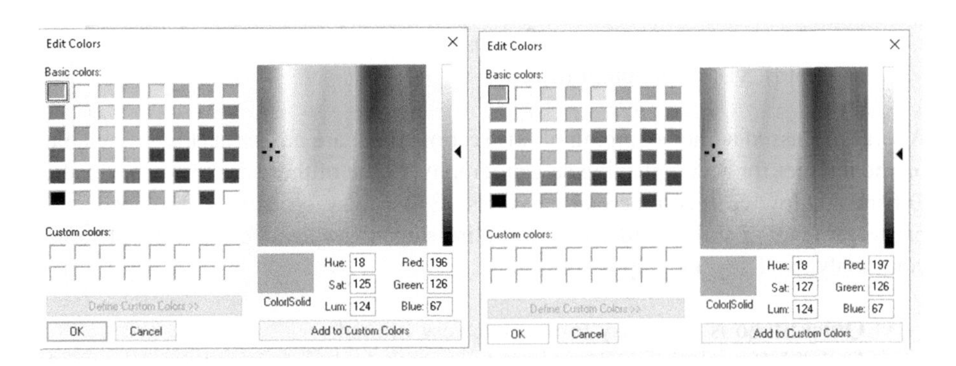

FIGURE 10.2 Changing 1 bit.

FIGURE 10.3 Two dogs after 1 bit changed.

- **Payload** is the information to be covertly communicated. In other words, it is the message you want to hide.
- The **carrier** (or carrier file) is the signal, stream, or file in which the payload is hidden.

- The **channel** is the type of medium used. This may be a passive channel, such as photos, video, or sound files.

As was discussed at the beginning of this section, there are numerous tools available on the internet for accomplishing steganography. Some of these are free downloads. Others are low-cost commercial products. Either way, one does not need any particular level of skill in order to accomplish steganography. Here is a list of some commonly used steganography tools.

- **QuickStego** is very easy to use, but very limited.
- **InvisibleSecrets** is much more robust, with both a free and a commercial version.
- **MP3Stego** hides a payload in MP3 files.
- **Stealth Files 4** works with sound files, video files, and image files.
- **StegVideo** hides data in a video sequence.
- **DeepSound** hides data in sound files.

As you can see, some of these are meant to hide data in sound files not images. You can also see software for hiding data in video files. Any file can be used to hide data. However, the larger the file is, the easier it is to hide something in it.

Regardless of the method uses or the carrier file, it is important to be able to detect steganographically hidden data. This is done via a process called steganalysis. Steganalysis is the process of analyzing a file or files for hidden content. This process can be time consuming and is not guaranteed. So, a good first step is determining if a given file is a likely candidate to have something steganographically hidden in it.

The first is examining metadata about a given image or sound file. Two pieces of data are of greatest interest: the created date and the last-modified date. The created date is not when the file was originally created, but rather when it was created *on that device*. If you download an mp4 file, then the date/time you download it to your computer will be the created date. The last-modified date indicates when it was last changed. Most people do not modify music files; they usually just download music, perhaps from Google Play or Apple Music. The last-modified date and created date will either be the same, or the last-modified date will be older. If the last-modified date is newer, this indicates the file has been modified since it was downloaded. If that person has music editing software and is known to mix music, then that explains the date discrepancy. However, if that is not the case, then steganography should be considered.

It is also possible to detect possible steganography via file size. Many steganography tools don't do a really good job of steganography and bloat the target file's size. This means seeing a file with an inappropriate size could indicate possible steganography. If a suspect has a collection of 100 vacation pictures, all of which are roughly 2 megabytes in size, but one is 4 megabytes in size, it may be worthwhile to analyze that image.

Steganalysis can be a difficult task. There are tools to do this. Once you have identified files you think might have steganographically hidden data, place those files on your forensic workstation for further analysis. A common method for detecting LSB steganography is to examine close-color pairs. Close-color pairs consist of two colors that have binary values that differ only in the LSB. If this is seen too frequently in a given file, it can indicate that steganographically hidden messages may be present. The raw quick pair method is a common way of checking for steganography. It is based on statistics of the numbers of unique colors and close-color pairs in a 24-bit image. Basically, it performs a quick analysis to determine if there are more close-color pairs than would be expected.

Another option uses the chi-square method from statistics. Chi-square analysis calculates the average LSB and builds a table of frequencies and a second table with pairs of values. Then it performs a chi-square test on these two tables. Essentially, it measures the theoretical versus the calculated population difference. When analyzing audio files, you can use steganalysis, which involves examining noise distortion in the carrier file. Noise distortion could indicate the presence of a hidden signal.

There are several free or inexpensive tools for detecting steganography:

- McAfee has an online steganography detection tool: https://www.mcafee.com/enterprise/en-us/downloads/free-tools/steganography.html
- StegDetect: https://github.com/abeluck/stegdetect
- Steg Secret is another tool: http://stegsecret.sourceforge.net/ has fewer limitations than StegDetect: http://www.spy-hunter.com/stegspydownload.htm

In order to better understand steganography, and to see how easy it is for suspects to hide data, we will use a few steganographic tools. The next few sections provide more details and a demonstration of a few widely used steganographic tools.

InvisibleSecrets

One very flexible tool for steganography is InvisibleSecrets. This tool is very inexpensive, and a free trial version is available. It is also easy to use. You can download InvisibleSecrets trial version from https://www.east-tec.com/invisiblesecrets/download/.

First, you must choose whether you want to hide a file or extract a hidden file.

The first step after launching the software is to select Hide Files. This is shown in Figure 10.4.

Once you have selected Hide Files, the wizard will prompt you to select a carrier file and the file you wish to hide. This is shown in Figure 10.5.

Then you simply continue through the wizard making choices. For example, you can choose to add a password, or even encrypt the file in addition to hiding it with steganography. When you are done, the tool will produce a new image with the data you added hidden in the image. This is very easy to use. The ease of use is one reason all forensic analysts should be familiar with steganography. It is quite likely that you will have a case that involves steganographically hidden data.

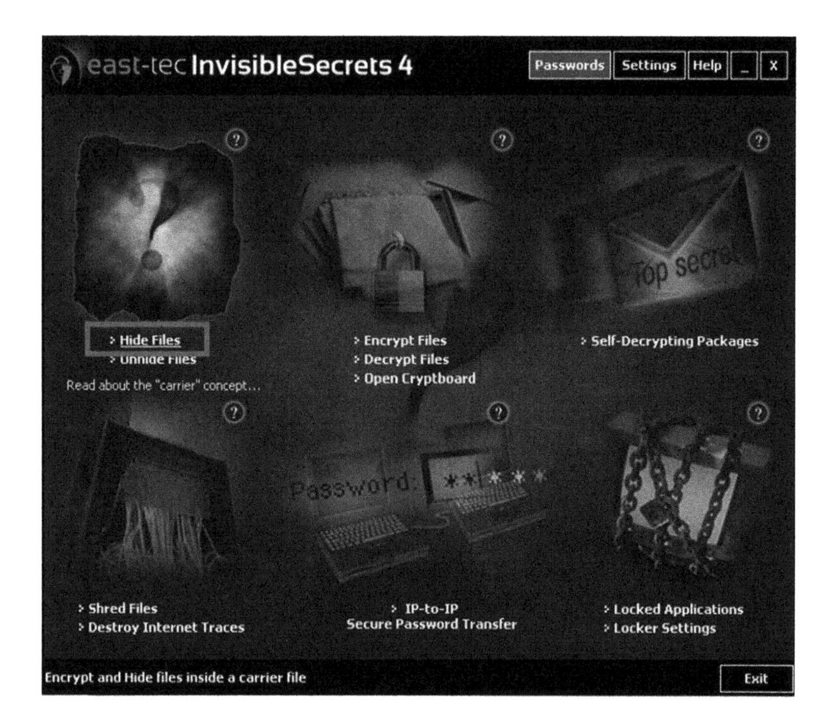

FIGURE 10.4 InvisibleSecrets initial screen.

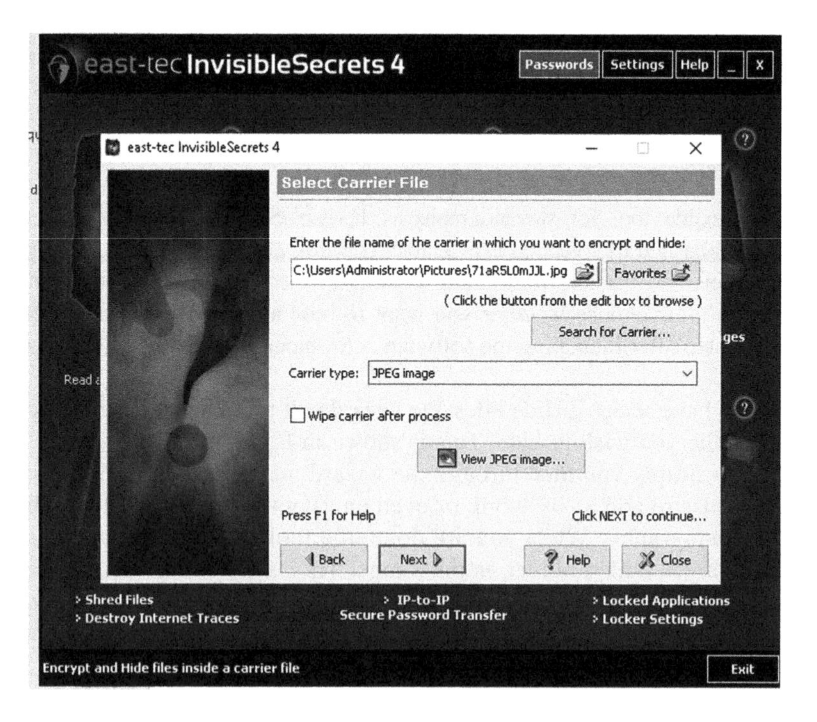

FIGURE 10.5 InvisibleSecrets wizard.

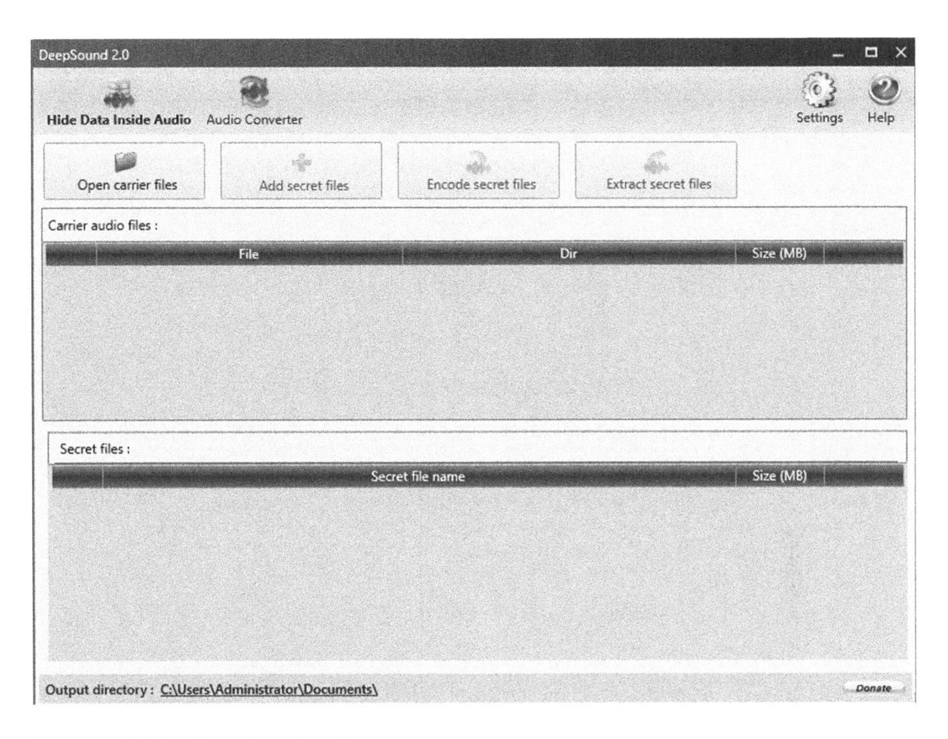

FIGURE 10.6 DeepSound main screen.

DeepSound

This is another tool for hiding data into sound files. It is a free download from https://deepsound.en.uptodown.com/windows. This tool is quite easy to use. The initial screen is shown in Figure 10.6.

The tool is quite simple to use. You first select the carrier file, then add 1 or more files you wish to hide, and then click encode secret files (Figure 10.7).

For this demonstration I took a syllabus for a course I teach and hid it in a concerto by Tchaikovsky. This should illustrate again, how easy it is for anyone, with software freely available on the internet, to use steganography. There are many other tools available. A simple web search for "free steganography tool" will confirm that.

CRYPTOGRAPHY

Modern cryptography is separated into two distinct groups: symmetric cryptography and asymmetric cryptography. Symmetric cryptography uses the same key to encrypt and decrypt the plaintext, while asymmetric cryptography uses different keys to encrypt and decrypt the message. Symmetric cryptography is often used for file and drive encryption. Asymmetric cryptography is used for exchanging symmetric keys, encrypting emails, or digitally signing messages. Since both iOS and Android devices employ encryption, it is important that you have at least some idea of how encryption works.

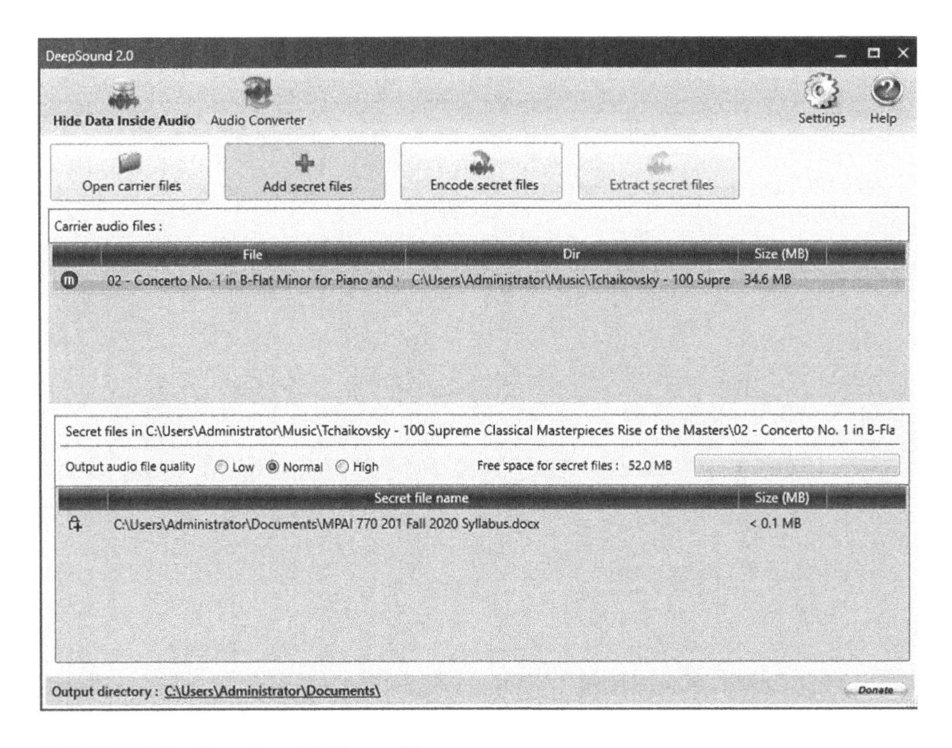

FIGURE 10.7 DeepSound hiding a file.

During our discussion of cryptography, we will look at the actual steps for four algorithms. Two symmetric algorithms and two asymmetric algorithms will be showcased. If you struggle a bit with the actual algorithms, do not be overly concerned. The goal of this chapter is not to try and make you a cryptographer. Rather it is to give you a solid understanding of how cryptography works, so you have some idea of how it might be possible to circumvent security on a mobile device.

Symmetric Cryptography

Symmetric cryptography uses the same key to encrypt the message as it does to decrypt. Symmetric algorithms are usually faster than asymmetric, and just as secure with smaller key sizes. Modern symmetric algorithms depend on a series of steps that involve substitution and transposition. Substitution is about changing bits of input text for different bits of output text. Transposition is the swapping of blocks of ciphertext. For example, if you have the text "I like ice cream," you could transpose or swap every three-letter sequence (or block) with the next and get:

"ikeI l creiceam"

The methods used in modern symmetric algorithms are certainly more complex than this. They also usually have many rounds of substitution and transposition. Another facet of modern symmetric ciphers is the key schedule algorithm. Modern symmetric algorithms such as AES, Blowfish, Serpent, etc. take the cryptographic key that was exchanged between the two parties and make minor adjustments to it each round. This is done according to a clearly defined algorithm called a key scheduling algorithm. This way, each round is not only applying substitution and transposition, but is usually slightly different round key.

There are two types of symmetric algorithms: block ciphers and stream ciphers. A block cipher literally encrypts the data in blocks; 64-bit blocks are quite common, although some algorithms (like AES) use larger blocks. For example, AES uses a 128-bit block. Stream ciphers encrypt the data as a stream, one bit at a time.

There are a few basic facts that are commonly applicable to all block ciphers. Assuming the algorithm is mathematically sound, then the following is true:

- Larger block sizes increase security.
- Larger key sizes increase security against brute-force attack methods.
- If the round function is secure, then more rounds increase security to a point.

The issue is the algorithm itself. If the algorithm is not sound, then it does not matter how many rounds, how large a block size, or how large a key is used, it won't be secure.

DES

The Data Encryption Standard is a fundamental algorithm to study in cryptography. Though it is no longer considered secure due to its small key size, the basic structure is used for several other algorithms. It was also the first publicly disclosed modern symmetric cipher. It was selected by the National Bureau of Standards as an official Federal Information Processing Standard (FIPS) for the United States in 1976. These facts make this a good place to begin studying symmetric ciphers.

DES uses a 56-bit cipher key applied to a 64-bit block. There is actually a 64-bit key, but one bit of every bite is actually used for error correction, leaving just 56 bits for actual key operations.

DES is a Feistel cipher with 16 rounds and a 48-bit round key for each round. A Feistel cipher is one that splits each block of plain text into two halves. One half is put through a round function that performs the transposition and substitution. Then the two halves are swapped. This is named after Horst Feistel, the inventor of the Feistel function and the primary inventor of DES.

The first issue to address, is the key schedule. Recall earlier we described the key schedule as an algorithm that generates a slightly different key each round, derived from the key the two parties exchanged. Each symmetric cipher has its own key schedule. The DES key schedule is rather simple. To generate the round keys, the 56-bit key is split into two 28-bit halves and those halves are circularly shifted after each round by one or two bits. This will provide a different sub key each round.

TABLE 10.1
DES Key Schedule Algorithm

Round	How Far to Shift to The Left
1	1
2	1
3	2
4	2
5	2
6	2
7	2
8	2
9	1
10	2
11	2
12	2
13	2
14	2
15	2
16	1

During the round key generation portion of the algorithm each round, the two halves of the original cipher key are shifted by a specified amount. The shifts are shown in Table 10.1.

Recall that DES splits each 64-bit plain text block into 2 halves, each 32 bits. One of these halves is put through the round function. Given the round key is 48 bits, there is a mismatch between the size of the round key and the 32-bit half block of plain text it will be applied to. Recall that the two halves are each 32 bit. You cannot really XOR a 48-bit round key with a 32-bit ½ block, unless you simply ignore 16 bits of the round key. The 32-bit half needs to be expanded to 48 bits before it is XORd with the round key. This is accomplished by replicating some bits so that the 32-bit half becomes 48 bits. Recall the XOR operation was discussed in chapter 1.

This expansion process is relatively simple. The 32 bits that is to be expanded is broken into 8 sections, each 4-bits long. The bits on each end of these 4-bit sections are duplicated, making 6-bit sections. Having 8 sections, each six bits long is 48 bits. That is the same size as the round key. The next step is to exclusively or (XOR) the round key with the expanded 48 bits. That is the extent of the round key being used in each round. It is not used any further in the algorithm, and on the next round another 48-bit round key will be derived from the two 28-bit halves of the 56-bit cipher key.

Now we have the 48-bit output of the XOR operation. That is now split into 8 sections of 6 bits each. For the rest of this explanation, we will focus on just one of those 6-bit sections, but keep in mind that the same process is done to all 8 sections.

The six-bit section is used as the input to an s-box. An s-box is a table that takes input and produces an output based on that input. In other words, it is a substitution box that substitutes new values for the input. The s-box's used in DES are published, the first of which is shown in Figure 10.8.

Notice this is simply a lookup table. The two bits on either end are shown in the left-hand column and the four bits in the middle are shown in the top row. They are matched and the resulting value is the output of the s-box. As an example, consider the input is 111000. Therefore, you find 1yyyy0 on the left and x1100x on the top. The resulting value is in decimal or 0011 in binary.

Recall earlier that the original plaintext was divided into 4-bit sections and the end bits expanded to create 6-bit sections. This is why truncating the bits on either end, moving from 6-bt inputs to the s-box to 4-bit outputs does not lose any information.

At the end of this you have produced 32-bits that are the output of the round function. Then in keeping with the Feistel structure, they get XOR'd with the 32 bits that were not input into the round function, and the two halves are swapped. DES is a 16 round Feistel cipher, meaning this process is repeated 16 times.

This provides a general outline of how DES works. We have left off a few items, such as an initial permutation. However, this description is sufficient for our purposes. The concept is for you, the forensic examiner, to have a general knowledge of how symmetric cryptography works.

For a thorough discussion of the technical details of DES refer to the actual government standard documentation: U.S. DEPARTMENT OF COMMERCE/ National Institute of Standards and Technology FIPS Publication 46-3 http:// csrc.nist.gov/publications/fips/fips46-3/fips46-3.pdf For a less technical but interesting animation of DES you may try http://kathrynneugent.com/animation.html

AES

It eventually became apparent that DES was not going to be secure enough for future use. This was due to advances in computing power that rendered brute force attacks on DES very practical. The National Institute of Standards initiated a project to identify a new standard. The Rijndael block cipher was ultimately chosen as a replacement for DES. The standardized version of Rijndael is called Advanced Encryption Standard (AES). AES is designated as Federal Information Processing Standard (FIPS) 197.

AES can have three different key sizes, they are:128, 192, or 256 bits. The three different implementations of AES are referred to as AES 128, AES 192, and AES 256. The block size, however, is always 128-bit. It should be noted that the original Rijndael cipher allowed for variable block and key sizes in 32-bit increments. It should be noted that the algorithm, the Rijndael algorithm supports other key and block sizes. Rijndael supports key and block sizes of 128, 160, 192, 224, and

	x0000x	x0001x	x0010x	x0011x	x0100x	x0101x	x0110x	x0111x	x1000x	x1001x	x1010x	x1011x	x1100x	x1101x	x1110x	x1111x
0yyyy0	14	4	13	1	2	15	11	8	3	10	6	12	5	9	0	7
0yyyy1	0	15	7	4	14	2	13	1	10	6	12	11	9	5	3	8
1yyyy0	4	1	14	8	13	6	2	11	15	12	9	7	3	10	5	0
1yyyy1	15	12	8	2	4	9	1	7	5	11	3	14	10	0	6	13

FIGURE 10.8 The first DES s-box.

11011001	01110010	10110000	11101010
01011111	00011001	11011001	10011001
10011100	11011101	00011001	11111101
11011001	10001001	11011001	10001001

FIGURE 10.9 The Rijndael matrix.

256 bit. However, the AES standard specifies a block size of only 128 bits and key sizes of 128, 192, and 256 bit.

Rijndael uses a substitution-permutation matrix rather than a Feistel network. The Rijndael cipher works by first putting the 128-bit block of plain text into a 4-byte X 4-byte matrix. This matrix is termed the state and will change as the algorithm proceeds through its steps. Thus, the first step is to convert the plain text block into binary, then put it into a matrix as shown in Figure 10.9.

Rijndael Steps

The algorithm consists of a few relatively simple steps that are used during various rounds. The steps are described here:

AddRoundKey – each byte of the state is combined with the round key using bitwise xor. This is where Rijndael applies the round key generated from the key schedule. This key schedule is different from the DES key schedule but serves the same general purpose.

SubBytes – a non-linear substitution step where each byte is replaced with another according to a lookup table. This is where the contents of the matrix are put through the s-boxes. AES has a single s-box.

ShiftRows – a transposition step where each row of the state is shifted cyclically a certain number of steps. In this step the first row is left unchanged. Every byte in the second row is shifted one byte to the left (with the far left wrapping around). Every byte of the third row is shifted two to the left, and every byte of the fourth row is shifted three to the left (again with wrapping around). This is shown in Figure 10.10.

Initial State

1a	1b	1c	1d
2a	2b	2c	2d
3a	3b	3c	3d
4a	4b	4c	4d

After Shift Rows

1a	1b	1c	1d
2b	2c	2a	2a
3c	3d	3a	3b
4d	4a	4b	4c

FIGURE 10.10 Shift rows.

Notice that in Figure 10.10 the bytes are simply labeled by their row then a letter, for example, 1a, 1b, 1c, 1d.

MixColumns – a mixing operation which operates on the columns of the state, combining the four bytes in each column. In the MixColumns step, each column of the state is multiplied with a fixed polynomial. Each column in the state (remember the matrix we are working with) is treated as a polynomial within the Galois Field (2^8). The result is multiplied with a fixed polynomial $c(x) = 3x^3 + x^2 + x + 2$ modulo $x^4 + 1$. The mathematics of the MixColumns step are a bit beyond the scope of this current chapter. Fortunately, for our purposes, you don't need a deep understanding of that mathematics.

Rijndael Outline

The previously described steps are executed in a particular order in the Rijndael cipher. Different key sizes of AES have different numbers of rounds. For 128-bit keys, there are 10 rounds. For 192-bit keys there are 12 rounds. For 256-bit keys there are 14 rounds.

Key Expansion – The first step is that the round keys are derived from the cipher key using Rijndael's key schedule.

Initial Round

This initial round will only execute the addroundkey step. This is simply XOR'ing with the round key. This initial round is executed once, then the subsequent rounds will be executed.

Rounds

This phase of the algorithm executes several steps, in the following order:

SubBytes
ShiftRows
MixColumns
AddRoundKey

Final Round

This round has everything the rounds phase has, except no mix columns.

SubBytes
ShiftRows
AddRoundKey

In the AddRoundKey step, the subkey is xord with the state. For each round, a subkey is derived from the main key using Rijndael's key schedule; each subkey is the same size as the state.

This is just a general overview of AES, but sufficient for our purposes. The main item for you to understand is that AES is a robust algorithm. You are extremely unlikely to break the AES used by both iOS and Android devices. However, there are still ways to get into a mobile device. We will discuss those later in this chapter.

Improving Symmetric Ciphers

There have been multiple methods developed to make symmetric ciphers a bit more secure. One that is seen a great deal with mobile devices is CBC or cipher block chaining. What cipher block chaining does is take the output of each block, and xor it with the next block of plain text. Consider DES with 64-bit blocks. After the first block has been through 16 rounds of DES, you have a 64-bit block of ciphertext. And that will be produced as part of the total ciphertext. However, it will also be taken and xor'd with the next 64-bit block of plaintext, before that plaintext is encrypted. The outcome of this process is that even if every single 64-bt block of plaintext were exactly identical, the ciphertext produced would be quite different. This is a simple mechanism that can improve the security of any block cipher.

Asymmetric Cryptography

Earlier in this chapter, it was stated that symmetric cryptography is faster than asymmetric and just as secure with a smaller key. That might lead one to question why we have asymmetric cryptography at all. The reason is simple: key exchange. If asymmetric cryptography did not exist, then we would have a problem of how to exchange keys. The only secure mechanism would be to physically meet with the other party and exchange keys on some device such as a USB. That would render all ecommerce, secure email, and similar technology nonexistent. Thus, asymmetric cryptography has an important role to play in secure communications.

In this section we will review very commonly used asymmetric algorithms. The complete algorithms will be presented. If you have a limited math background, this might present some difficulties for you. It is not critical that you understand all the steps of these algorithms. Rather it is necessary for you to have a general working knowledge of these algorithms.

RSA

RSA may be the most widely used asymmetric algorithm. The algorithm was first publicly disclosed in 1977 by Ron Rivest, Adi Shamir, and Leonard Adleman. The name RSA is derived from the first letter of each mathematician's last name. The algorithm is based on prime numbers and the difficulty of factoring a large number into its prime factors.

First to create the public and private key pair you start by generating two large random primes, p and q, of approximately equal size (Hinek, 2009).[1] You will need to select two numbers so that when multiplied together the product will be the size you want (i.e., 2048 bits, 4096 bits, etc.)

Next, multiply p and q to get n.

Let $n = pq$

The next step is to multiply Euler's Totient for each of these primes.

If the term Euler's totient is new to you, a brief explanation is provided here. Leonard Euler was a mathematician who studied integers in the 1700s. One of the issues he explored was integers which are co-prime. Two numbers are considered co-prime if they have no common factors. For example, if the original number is 8, then 8 and 9 would be co-prime. 8's factors are 2 and 4, 9's factors are 3. Euler conducted further study of co-prime integers. Given a number X, how many numbers smaller than X are co-prime to X. We call that number Euler's totient, or just the totient. It just so happens that for prime numbers, this is always the number minus 1. For example, 7 has 6 numbers that are co-prime to it.

When you multiply two primes together, you get a composite number. And there is no easy way to determine the Euler's totient of a composite number. Euler found that if you multiply any two prime numbers together, the Euler's totient of that product is the Euler's totient of each prime multiplied together. So, our next step is:

Let $m = (p - 1)(q - 1)$

So, m, is the Euler's totient of n.

Now we are going to select another number, we will call this number e. We want to pick e so that it is co-prime to m. Frequently a prime number is chosen for e. That way if e does not evenly divide m, then we are confident that e and m are co-prime, as e does not have any factors to consider. Many RSA implementations use $e = 2^{16} + 1 = 65537$. This is considered large enough to be effective but small enough to still be fast.

At this point you have almost completed generating the key. Now we just find a number d that when multiplied by e and modulo m would yield a 1 (note: modulo means to divide two numbers and return the remainder. For example, 8 modulo 3 would be 2). In other words:

Find d, such that de % $m \equiv 1$

Now you will publish e and n as the public key.

Keep d as the secret key. To encrypt you simply take your message raised to the e power and modulo n.

Ciphertext = Message^{e} mod n

To decrypt you take the cipher text, raise it to the d power modulo n.

Plaintext = Ciphertext^{d} mod n

To make this clearer, let us examine an example. To make the math easy to follow we will use small integers in this example. (note this example is from Wikipedia):

Choose two distinct prime numbers, such as $p = 61$ and $q = 53$.

Compute $n = pq$ giving $n = 61 \cdot 53 = 3233$.

Compute the totient of the product as $\varphi(n) = (p - 1)(q - 1)$ giving $\varphi(3233) = (61 - 1)(53 - 1) = 3120$.

Choose any number $1 < e < 3120$ that is co-prime to 3120. Choosing a prime number for e leaves us only to check that e is not a divisor of 3120. Let $e = 17$.

Compute d, the modular multiplicative inverse of yielding $d = 2753$.

The public key is ($n = 3233$, $e = 17$). For a padded plaintext message m, the encryption function is m^{17} (mod 3233).

The private key is ($n = 3233$, $d = 2753$). For an encrypted ciphertext c, the decryption function is c^{2753} (mod 3233).

For those readers new to RSA, or new to cryptography in general, it might be helpful to see one more example, with even smaller numbers.

1. Select primes: $p = 17$ & $q = 11$
2. Compute $n = pq = 17 \times 11 = 187$
3. Compute $\varnothing(n) = (p-1)(q-1) = 16 \times 10 = 160$
4. Select e: gcd(e,160) = 1; choose $e = 7$
5. Determine d: $de = 1 \mod 160$ and $d < 160$ Value is d = 23 since $23 \times 7 = 161 = 10 \times 160 + 1$
6. Publish public key (7 and 187)
7. Keep secret private key 23

Now let us apply this second example of RSA key generation to actually encrypting something. For some reason you have decided to encrypt the number 3. Thus, we will use the number 3 as the plain text. Remember $e = 7$, $d = 23$, and $n = 187$. So, we will use the recipients public key which is e and n.

Ciphertext = Plaintexte mod n or
Ciphertext = 37 mod 187
Ciphertext = 2187 mod 187
Ciphertext = 130

When the recipient receives this ciphertext, he or she will use their secret key to decrypt it as follows:

Plaintext = Ciphertext d mod n
Plaintext = 13023 mod 187
Plaintext = 4.1753905413413116367045797e + 48 mod 187
Plaintext = 3

As you can see it works, even if there is not really any particular reason to encrypt the number three. As you can see the RSA algorithm actually works to encrypt and decrypt. As of this writing, many RSA implementations use key sizes of 2048 or larger. For interested readers, there is an excellent YouTube video that explains RSA for novices https://www.youtube.com/watch?v=wXB-V_Keiu8&t=254s.

Diffie–Hellman

Diffie–Hellman is a cryptographic protocol that allows two parties to establish a shared key over an insecure channel. This algorithm was first publicly disclosed by Whitfield Diffie and Martin Hellman in 1976.

The Algorithm has two parameters called p and g. Parameter p is a prime number and parameter g (usually called a generator) is an integer less than p, with the following property: for every number n between 1 and p-1 inclusive, there is a power k of g such that $n = g^k \mod p$. Let us revisit our old friends Alice and Bob to illustrate this:

> Alice generates a random private value a and Bob generates a random private value b. Both a and b are drawn from the set of integers

FIGURE 10.11 Diffie–Hellman.

They derive their public values using parameters p and g and their private values. Alice's public value is g^a mod p and Bob's public value is g^b mod p. They exchange their public values.
Alice computes $g^{ab} = (g^b)^a$ mod p, and Bob computes $g^{ba} = (g^a)^b$ mod p.
Since $g^{ab} = g^{ba} = k$, Alice and Bob now have a shared secret key k.

This process is shown in Figure 10.11.

Again, do not be overly concerned if the math is somewhat difficult for you. The goal is for you to understand key exchange algorithms. There is an excellent video on YouTube that explains Diffie–Hellman with colors https://www.youtube.com/watch?v=YEBfamv-_do&t=15s.

Mobile Device Specific Cryptography

Both iOS and Android use AES 256-bit encryption for encrypting the drive. However, there are specific cryptographic algorithms used in the communication between the device and the cell phone tower. These were discussed briefly in chapter 1. There were originally 7 algorithms specified for GSM use. A5/1 is a stream cipher designated for communications between the mobile device and the tower. Eventually A5/1 was found to have weaknesses to certain specific attacks. The first published attack was proposed by Ross Anderson, an eminent cryptography, in 1994. A5/2 was a deliberately weakened version of A5/1 for use in certain expert situations.

A5/3 is also known by the algorithm name KASUMI, was the replacement for A5/1. The name KASUMI comes from the Japanese word for mist, because KASUMI is an improvement on the MISTY1 algorithm. KASUMI is a block cipher with a 128-bit key. It is an 8 round Feistel cipher. Recall Feistel ciphers divide the input blocks in half, putting one half through a round function, then using the xor operation on one half, then swapping the halves each round.

LTE uses A5/4 also known as the SNOW algorithm. SNOW supports either 128 or 256-bit keys. The original SNOW had some security issues, but version 2 of the algorithm solved those weaknesses.

LTE actually defines three algorithms:

- 128-EEA1/EIA1: This is based on SNOW.
- 128-EEA2/EIA2: This uses AES with CMAC for integrity.
- 128-EEA3/EIA3: This uses the Zu Chongzhi (ZUC) stream cipher.

The various Ax algorithms are used for different purposes. A3 is used for authentication. Specifically, to authenticate a mobile station to the network. A5 is used to encrypt transmissions. And A8 is used to generate a session key to be used by A5 for encryption between the base transceiver station (BTS) and the mobile station.

A5/3 Block cipher used in UMTS, GSM, and others. It was designed explicitly for the 3GPP Standard by Security Algorithms Group of Experts (SAGE), a part of the European standards body ETSI.

Recall in chapter 1 we discussed UMTS authentication. While the algorithms have been changed, the basic process is much the same. Authentication begins with a pre-shared secret key that is stored in the Authentication Center (AuC) of the network, and in the USIM of the device. The pre-shared key will be used to help create a cipher and integrity key. The cipher and integrity key are new for each session.

More specifically, the AuC and the USIM share some specific components:

- The secret key specific to each subscriber
- Authentication algorithms that consist of authentication functions (f1, f1*, and f2)
- Key generation functions (f3, f4, f5, f5*).

The AuC has a random number generator and a sequence number generator. The USIM has an algorithm to assure that sequence numbers received from the AuC are new. The specific parameters used in UMTS authentication are given here:

- K = Subscriber authentication key (128 bit)
- RAND = User authentication challenge (128 bit)
- SQN = Sequence number (48 bit)
- AMF = Authentication management field (16 bit)
- MAC = $f1_K$ (SQN || RAND || AMF) (64 bit)
- (X)RES = $f2_K$ (RAND)

= (Expected) user response (32 – 128 bit)

- CK = $f3_K$ (RAND) = Cipher key (128 bit)
- IK = $f4_K$ (RAND) = Integrity key (128 bit)
- AK = $f5_K$ (RAND) = Anonymity key (48 bit)
- AUTN = SQN⊕AK|| AMF||MAC (128 bit)

The ⊕ symbol denotes the exclusive or (XOR) process discussed earlier in this chapter.

The primary cryptographic algorithm used is the UMTS Encryption Algorithm (UEA1). This algorithm makes use of the KASUMI block cipher. KASUMI was specifically designed for the 3GPP standard.

In February 2019, Google unveiled Adiantum, an encryption cipher designed primarily for use on devices that do not have hardware-accelerated support for the Advanced Encryption Standard (AES), such as low-end devices. This is for device encryption, not encrypting voice transmissions.

Adiantum is a cipher construction for disk encryption, which uses the ChaCha and Advanced Encryption Standard (AES) ciphers, and Poly1305 cryptographic message authentication code (MAC). ChaCha is a variant of the Salsa stream cipher.

In 2013, Mouha and Preneel published a proof that 15 rounds of Salsa20 was 128-bit secure against differential cryptanalysis. Specifically, it has no differential characteristic with higher probability than $2 - 130$, so differential cryptanalysis would be more difficult than 128-bit key exhaustion.

Cryptographic Hash

You have undoubtably heard of cryptographic hashes before. A cryptographic hash is a special type of algorithm. In order to be a cryptographic hash function, an algorithm needs to have three properties. The first property is that the function is one way. That means it cannot be "unhashed." Now this may seem a bit odd at first. An algorithm that is not reversible? Not simply that it is difficult to reverse, but that it is literally impossible to reverse. Yes, that is exactly what I mean. Much like trying to take a scrambled egg and unscramble it and put it back in the eggshell, it is just not possible. When we examine specific hashing algorithms later in this chapter, the reason why a cryptographic hash is irreversible should become very clear.

The second property that any cryptographic has must have is that a variable length input produces a fixed-length output. That means that no matter what size of input you have, you will get the same size output. Each particular cryptographic hash algorithm has a specific size output. For example, SHA-1 produces a 160-bit hash. It does not matter whether you input 1 byte or 1 terabyte, you get out 160 bits.

How do you get fixed-length output regardless of the size of the input? Different algorithms will each use their own specific approach but in general it involves compressing all the data into a block of a specific size. If the input is smaller than the block, then pad it.

Cryptographic hashes are used for message integrity, and for password storage. Most operating systems, including mobile operating systems, store passwords as a hash. Later in this chapter, we will discuss a method widely used to retrieve passwords from cryptographic hashes.

When used for integrity and authentication, hashes are often enhanced with the addition of a key. An HMAC is a traditional hash, but the output is then xor'd with a randomly generated key. A CMAC is a symmetric cipher using cipher block chaining so that only the final block is produced and used as a hash.

PASSWORD CRACKING

A number of tools can aid in cracking passwords and encrypted data. Remember that if the encryption was implemented correctly and is strong, you may not be able to crack it. But passwords can often be cracked (encrypted information less often). It is also possible to obtain keys or copies of information before encryption via a number of nontechnical means that fall in the category of social engineering, which includes going through the trash, also known as dumpster diving; lying to a person to obtain the keys, passwords, phrases, or unencrypted information; or even getting a job at the target company and stealing the desired information.

In 1980, Martin Hellman described a cryptanalytic time-memory tradeoff, which reduces the time of cryptanalysis by using precalculated data stored in memory. Essentially, these types of password crackers work with precalculated hashes of all passwords available within a certain character space, be that a–z or a–zA–z or a–zA–Z0–9, etc. These files are called rainbow tables because they contain every letter combination "under the rainbow." They are particularly useful when trying to crack hashes. Because a hash is a one-way function, the way to break it is to attempt to find a match. The attacker takes the hashed value and searches the rainbow tables seeking a match to the hash. If one is found, then the original text for the hash is found.

Various tools claim to be able to crack passwords for iPhone and Android. As you can probably imagine, their efficacy varies. A great deal depends on the complexity of the password or PIN used to secure the phone.

4uKey https://www.tenorshare.net/

My iPhone Wiper https://www.imyfone.com/unlock-iphone/

Nyuki Android Cracker https://www.silensec.com/downloads-menu/software/androidlockcracker

You may find some limited success with these tools. However, it should be obvious that unlocking a phone is not going to work most of the time. If it did, then the phone security would be useless. It is a frustrating aspect of mobile forensics that you sometimes cannot get into a suspects phone. General techniques that may have success include:

- For iPhone's get the Apple ID and password from the suspect's computer.
- Look to see if the suspect uses the same password or pin in multiple places. It is generally easier to extract passwords from Windows. If you can extract passwords from the suspect's Windows computer these may provide insight into the mobile device password or pin.
- Look for patterns with the suspect. This can include things such as a strong affinity for a given sports team. This may provide some insight into the password or pin number for the phone.

GrayKey

GrayKey is a tool sold only to law enforcement that has had success breaking iOS device encryption. The product website is https://go.grayshift.com. GrayKey is not inexpensive, costing 15,000 to 30,000 U.S. dollars for a license. The time needed to break iOS encryption varies from several minutes to many hours and it is not guaranteed to be able to break the encryption. Particularly on devices running iOS 12 and later, GrayKey has had some limitations. Reports from various law enforcement officers with direct experience are generally positive. Details of how the tool works are unavailable. GrayKey is quite strict in non-disclosure agreements and keeps information about its product confidential. In 2021 GrayKey announced it can now break encryption for Android devices.

SUMMARY

This chapter has introduced you to passwords, pins, steganography, and encryption. These are all methods that are used to prevent forensic examiners from extracting evidence from a suspect's computer. The chapter has also discussed possible mechanisms to circumvent these security features. However, it will often be the case that you will be unsuccessful in circumventing device security. That is why these devices use such security.

CHAPTER 10 ASSESSMENT

1. The most common method of steganography is what?
   a. Symmetric
   b. Asymmetric
   c. LSB
   d. Hashing
2. What is the term for the file you hide data in with steganography?
   a. Payload
   b. Carrier
   c. Channel
   d. Cover
3. What is a key schedule?
   a. An algorithm for generating round keys from the cipher key.
   b. A schedule for changing your symmetric key.
   c. A schedule for changing your asymmetric key.
   d. An algorithm for storing symmetric keys.
4. What is the most widely used asymmetric cryptographic algorithm today?
   a. DES
   b. AES
   c. DH
   d. RSA

5. Which of the following is the best description for a rainbow table?
   a. A table of keys to attempt brute force on a symmetric cipher.
   b. A table of keys to attempt brute force on an asymmetric cipher.
   c. A table of precomputed hashes.
   d. A table of known passwords for password cracking.

CHAPTER 10 LABS

LAB 10.1

Try one of the following tools to see if it can unlock an Android or iPhone (whichever you have access to. Note: If this is your phone or any phone actively in use (as opposed to a laboratory phone), make sure you do not factory reset the phone or erase data.

4uKey https://www.tenorshare.net/
My iPhone Wiper https://www.imyfone.com/unlock-iphone/
Nyuki Android Cracker https://www.silensec.com/downloads-menu/software/androidlockcracker
Dr. Fone password cracker https://drfone.wondershare.com/unlock/

NOTE

1 Hinek, M. J. (2009). Cryptanalysis of RSA and its variants. CRC press.

11 Legal Issues

INTRODUCTION

The previous chapters have covered a wide range of technical issues related to mobile phone forensics. However, there are also substantial legal issues. While the details may vary somewhat between jurisdictions, all courts have rules regarding admissibility of evidence and expert testimony. It is a serious mistake for a forensic examiner to not be aware of these issues. Understanding the legal issues may keep your findings from being excluded by a court. It is also true that following the legal guidelines can lead to better forensics.

RULES OF EVIDENCE

Rules of evidence govern whether, when, how, and why proof of a legal case can be placed before a judge or jury. A forensic examiner must have a working knowledge of the rules of evidence in the given type of court and jurisdiction. One example is the United States Federal Rules of Evidence (FRE) is a code of evidence law. The FRE governs the admission of facts by which parties in the U.S. federal court system may prove their cases. This means that these rules determine whether the data found will be presented to the jury or excluded.

The FRE provides guidelines for the authentication and identification of evidence for admissibility under Federal rules 901 and 902. The following is an excerpt from rule 901 of the FRE from Cornell University Law School) with the portions relevant to computer forensics shown:

(a) In General. To satisfy the requirement of authenticating or identifying an item of evidence, the proponent must produce evidence sufficient to support a finding that the item is what the proponent claims it is....

(1) Testimony of a Witness with Knowledge. Testimony that an item is what it is claimed to be.
(2) Comparison by an Expert Witness or the Trier of Fact. A comparison with an authenticated specimen by an expert witness or the trier of fact....
(3) Evidence About a Process or System. Evidence describing a process or system and showing that it produces an accurate result.

What this means is that the evidence presented must be authenticated. How can the court know that this is the hard drive seized from the suspect's computer? In many cases, adequate documentation can be the answer to addressing at least some of these issues. For example, the chain of custody is critical to authenticating evidence.

DOI: 10.1201/9781003118718-11

Description of Evidence		
Item #	**Quantity**	**Description of Item** (Model, Serial #, Condition, Marks, Scratches)

Chain of Custody				
Item #	**Date/Time**	**Released by** (Signature & ID#)	**Received by** (Signature & ID#)	**Comments/Location**

FIGURE 11.1 Generic chain of custody form.

There are differences in the format used, but the basic information required in a chain of custody form is the same in all jurisdictions. A sample chain of custody form is shown in Figure 11.1.

Again, different agencies will have variations, but the essentials are the same. The item, the data and time, who released and who received it. For example, if you receive an Android phone to forensically examine, you should record the date and time, who gave it to you, your own signature as receiving it, and details about the phone that uniquely identify it. Then everyone who handles that device should make an entry on the chain of custody form.

In addition to maintaining chain of custody records, the device itself must be stored securely. At an absolute minimum that requires a lockable cabinet with as few people as possible having access to it. And that is a bare minimum. It would be best to have a security camera and alarm system on a locked room that has the locked cabinet.

US Army Digital Evidence Storage

While you may not be employed with the United States Army, their rules of storing digital evidence can be a useful guideline for any forensic lab[1]:

1. "The evidence custodian should store digital media evidence in a dust-free, temperature- and humidity-controlled environment, whenever possible.
2. A person with digital media evidence will not store it near batteries, generators, electro-magnets, magnets, induction coils, unshielded microwave sources, or any material that generates static. NOTE: Vacuum cleaner motors

generate small electromagnetic fields that may alter, erase, and/or destroy digital media such as tapes.

3. A person with digital media evidence should not store such evidence in the same container with electronic devices. Some electronic devices contain batteries with sufficient strength to erase digital data over extended periods.
4. The evidence custodian should make periodic checks of digital media evidence in the evidence room to determine battery life of the item(s). There is a very high risk that all evidence contained in digital storage in these with appropriate chargers that can remain connected to uninterrupted power.
5. Where possible, the evidence custodian should store digital media evidence in a fire safe designed to safeguard items in heat in excess of 120 degrees Fahrenheit.
6. Where possible, the evidence custodian should not store digital media or devices in areas with sprinkler fire protection systems. If this is not possible, the evidence custodian should cover the media with waterproof material. The media should not be completely wrapped in waterproof material, because condensation can build and destroy the evidence.
7. The evidence custodian should not store digital media and devices in the same confined area with caustic chemicals (for example, acids, solvents, industrial strength cleaners, flammables). Exposure to fumes from such materials may cause surface erosion of media and loss of data."

Some of these may not be relevant to your lab. For example, you may not have an issue with caustic chemicals being nearby. However, these rules, in general, discuss securing the evidence against environmental threats. It is certainly possible for environmental issues to destroy evidence. The United States Army guidelines are quite extensive covering a wide range of possible evidence situations. Some of which are not pertinent to digital evidence. However, it is recommended that you peruse their guidelines to determine what is relevant to your lab.

Evidence Tracking

It is not uncommon for a forensic examiner to have multiple cases going simultaneously. In addition to that complexity, there are usually multiple examiners working in the same lab. This can make it quite easy to lose track of evidence. This issue is mitigated by using a tracking mechanism. There are four primary methods of tracking evidence: logs, software, barcode, and RFID. Each will be examined in the following subsections.

Logs

The most basic form of tracking is to maintain a log of evidence. This can be as simple as a document wherein each piece of evidence is tracked as it is removed from storage and returned. Figure 11.2 is a sample of such a log.

Obviously this is just an example, and you will likely see many variations in the field. The issue is that every time someone touches that evidence, there needs to be a log of what is occurring. Who touched the evidence, for what purpose, and when.

Case : 1111 Item: iPhone 11 pro Max Seriel Number XXX999222

Date Checked out	Time Checked out	Person Responsible	Reason	Time Checked in	Date Checked in
1/1//2021	n/a	Det. Smith	Initial check of evidence	9:12 am	1/1//2021
1/15/2021	11:14 am	Analyst Juan Perez	Initial Triage	2:11 pm	1/15/2021
1/16/2021	3:30 pm	Analyst Juan Perez	Exam with Cellebrite	5:45 pm	1/16/2021

FIGURE 11.2 Evidence tracking log.

It is critical to be able to account for the disposition of evidence from the moment it is seized to the moment it is presented in a court.

Software Tracking

Essentially software tracking is simply a software variation of the log. There are a variety of software tracking packages that will essentially take the place of a paper log. These packages simply make the maintaining and storing of logs more efficient. For example, these applications utilize databases, which will allow you to search for any item of interest. For example, you can search for all entries related to a specific device, date, or examiner. There are several software applications available for this purpose including:

ACISS Property and Evidence Management https://www.aciss.com/Solutions/Property-Evidence-Management-EMS

Erin 7 Management software https://erintechnology.com/

Evidence Tracker https://trackerproducts.com/evidence-software/

ASAP systems http://asapsystems.com/evidencetrac.php

There are other options, this is just an exemplary list. You should carefully review your agency's needs as well as budget in deciding which software is right for you.

Barcode

Barcode tracking essentially ties barcode technology to software tracking System details can vary from implementation to implementation but in general the system works as follows. Each piece of evidence has a barcode attached to it or to its container (such as an evidence bag). Then each analyst has their own barcode scanner. When a piece of evidence is accessed, its barcode is scanned. Then the system knows who accessed the evidence and when. This just makes software tracking easier and faster.

RFID

The ultimate in evidence tracking is RFID. Radio Frequency ID chips can be utilized to track evidence. These systems are implemented in much the same way as a barcode. The RFID chip is tracked throughout the lab, and movement is automatically logged into tracking software. The RFID chip is attached to the evidence, or the evidence container. Then at any given time the system knows the exact location of a given piece of evidence. As you might imagine, this can also be the most expensive solution.

EXPERT TESTIMONY

At some point evidence is presented in court. That requires that the forensic examiner testifies as to his or her examination and findings. Expert testimony is a bit different than eyewitness testimony. To begin with, the expert witness is not testifying as to events he or she saw. Instead, the expert is testifying as to their analysis and conclusions. There are federal rules in the United States that describe expert testimony.

U.S. Federal Rule 702 defines what an expert is and what expert testimony is:

> "A witness who is qualified as an expert by knowledge, skill, experience, training, or education may testify in the form of an opinion or otherwise if:
>
> a. the expert's scientific, technical, or other specialized knowledge will help the trier of fact to understand the evidence or to determine a fact in issue;
> b. the testimony is based on sufficient facts or data;
> c. the testimony is the product of reliable principles and methods; and
> d. the expert has reliably applied the principles and methods to the facts of the case"[2]

This definition of what constitutes expert testimony is quite useful. In addition to providing insight into how a court views such testimony, it can be a guide to how you conduct your forensic examinations. The first part begins with qualifications. Unfortunately, interest in digital forensics has exploded and lead to many entering the field without appropriate qualifications.

It would be best if one had formal education in computer science or a related field. But when that is not the case, it should at least be that the examiner has a solid knowledge of the relevant technology. That is one reason this book has included technical details well beyond simply using tools. A mobile forensic examiner should have a solid understanding of mobile device technologies.

Secondly is the testimony based on sufficient facts and is it the product of reliable principles and methods. This relates to how you, the examiner, conduct your analysis. You should only use well tested and confirmed tools and techniques. This will be expounded upon later when we discuss the Daubert standard. It should also be the case that you reliably applied those tools and techniques. Mobile technology is constantly changing. That requires forensic examiners to constantly be expanding and updating their skillset.

In the United States, federal courts apply the Daubert Standard to determining what expert testimony and evidence is allowed in court. This standard stems from a court case Daubert v Merrill Dow Chemical Inc. Prior to the Daubert case, the Frye v. United States case from 1923 defined the rules regarding admission of scientific evidence and testimony. The issue was that parties in the Daubert case argued that the earlier Frye standard did not take into account Rule 702 of the Federal Rules of Evidence, which we just examined.

The goals of the earlier Fry standard, Federal Rule 702, and the Daubert standard are all similar. Scientific testimony in court should be based on well-founded scientific principles, and the expert testifying must have properly applied those standards. The Daubert case simply refined the details of scientific testimony is evaluated. The Daubert standard is described as follows:

> Standard used by a trial judge to make a preliminary assessment of whether an expert's scientific testimony is based on reasoning or methodology that is scientifically valid and can properly be applied to the facts at issue. Under this standard, the factors that may be considered in determining whether the methodology is valid are: (1) whether the theory or technique in question can be and has been tested; (2) whether it has been subjected to peer review and publication; (3) its known or potential error rate; (4) the existence and maintenance of standards controlling its operation; and (5) whether it has attracted widespread acceptance within a relevant scientific community.

The practical result of the Daubert standard is that any mobile forensics technique must have been subjected to testing or peer-reviewed publication before it can be admissible in a court. The reason for these standards is that a judge may not have the requisite knowledge to distinguish between reliable scientific methods and junk science. Therefore, the onus is upon the relevant scientific community. That is why throughout this book it has been mentioned that tools and techniques have been vetted by the relevant scientific community.

The issue of admissibility of scientific evidence is by no means limited to the United States. All court jurisdictions have to address the problem of how to determine if scientific evidence is valid. And all courts face the same problem that the judge is unlikely to have relevant expertise to make that determination. The United Kingdom has a slightly different standard from Daubert (Crown Prosecution Service, 2019):

> There should be a sufficiently reliable scientific basis for the expert evidence, or it must be part of a body of knowledge or experience which is sufficiently organized or recognized to be accepted as a reliable body of knowledge or experience.
>
> The reliability of the opinion evidence will also take into account the methods used in reaching that opinion, such as validated laboratory techniques and technologies, and whether those processes are recognized as providing a sufficient scientific basis upon which the expert's conclusions can be reached. The expert must provide the court with the necessary scientific criteria against which to judge their conclusions.

While the wording is different than the United States Daubert standard, the general result is the same. Even if an expert is qualified by education and experience, the techniques that he or she applies must have had some validation in the relevant scientific community. The goal of both the United Kingdom standard and the United States standard is to ensure the expert testimony is based on sound scientific evidence.

WARRANTS

Beyond the issues of the forensic examiner's qualifications and methodology is the issue of permission. If you do not have permission from the legal owner of a piece of equipment, you may only perform an examination with a warrant. It is important to bear in mind that not only has the U.S. Supreme court held that a warrant is needed to seize property, but the court has also expounded upon what seizing means. According to the Supreme Court, a "'seizure' of property occurs when there is some meaningful interference with an individual's possessory interests in that property", United States v. Jacobsen, 466 U.S. 109, 113 (1984), and the Court has also characterized the interception of intangible communications as a seizure in the case of Berger v. New York, 388 U.S. 41, 59–60 (1967). Now that means that law enforcement need not take property in order for it to be considered seizure. Merely interfering with an individuals' access to his or her own property constitutes seizure.

And Berger v. New York extends that to communications. Now if law enforcements conduct does not violate a person's "reasonable expectation of privacy," then formally it does not constitute a Fourth Amendment "search" and no warrant is required. There have been many cases where the issue of reasonable expectation of privacy has been argued. This book does not offer legal advice. However, it is recommended that if you have any question about reasonable expectation of privacy, that you consult an attorney before proceeding with your examination.

In computer crime cases, two consent issues arise particularly often. The first issue involves exceeding the scope of consent. For example, when a person agrees to the search of a location, for example their apartment, does that consent authorize examination of a mobile phone at that address? Normally it would not. Again, this book does not offer legal advice. However, consent should be clear and specific.

The second issue involves who is the proper party to consent to a search? Can roommates, friends, and parents legally grant consent to a search of another person's computer files? These are all very critical questions that must be considered when searching a mobile device. In general courts have held that the actual owner of a property can grant consent. For example, a parent of a minor child can grant consent to search the living quarters and mobile phones. However, a roommate who shares rent can only grant consent to search shared living quarters.

ETHICS

Beyond the issues of legality are issues of ethics. The impact of digital forensics is extensive. Whether we are discussing civil cases or criminal, the effect is substantial. Forensic evidence and testimony frequently is a deciding factor in cases. This should indeed make you feel a bit nervous. Your findings can change the outcome of a criminal trial or a major lawsuit. It is critical that you do everything you can to behave ethically.

The American Academy of Forensic Science encompasses the entire breadth of forensics, not just cyber forensics. That includes fire forensics, medical forensics, financial forensics, etc. While the technical details of each type of investigation are quite diverse, the ethical challenges are remarkably similar. But due to the wide range of forensic sub disciplines, most of the AAFS ethical guidelines are necessarily broad. The actual ethical guidelines are listed here:

a. Every member and affiliate of the Academy shall refrain from exercising professional or personal conduct adverse to the best interests and objectives of the Academy. The objectives stated in the Preamble to these bylaws include: promoting education for and research in the forensic sciences, encouraging the study, improving the practice, elevating the standards, and advancing the cause of the forensic sciences.
b. No member or affiliate of the Academy shall materially misrepresent his or her education, training, experience, area of expertise, or membership status within the Academy.
c. No member or affiliate of the Academy shall materially misrepresent data or scientific principles upon which his or her conclusion or professional opinion is based.
d. No member or affiliate of the Academy shall issue public statements that appear to represent the position of the Academy without specific authority first obtained from the Board of Directors.

Items a and d are specific to members of the AAFS. However, items b and c are generally applicable to ethics whether you are a member or not. Item b basically admonishes you not to exaggerate. This is truly marvelous advice, though it may seem unnecessary. You might wonder if any expert would really exaggerate his or her credentials. The unfortunate answer is that yes some would. In fact, I have personally encountered incidents of this sort of behavior. For example, someone might have a legitimate degree, even a master's degree, but to enhance his or her resume they might get a doctorate degree from an unaccredited degree mill. Someone might claim certifications they don't have. The instances of forensic examiners exaggerating their credentials, or overstating the importance of their findings, are numerous.

Item c may seem even more odd. The fact is that in many expert reports, there may be dozens of citations. The people reading the report, particularly a lengthy one, might not check carefully every citation. I have personally prepared expert reports that were close to 300 pages long with many scores of citations. It would certainly be possible to exaggerate or interpret the conclusions of one or more of those citations in order to enhance one's position. However, that is unethical. And if the opposing side does check the reference, it will be quite damaging to your credibility.

One simple thing should aid you in making ethical decisions. Remember that you are not an advocate for either side. The attorneys are advocates. It is their job to try to win the case. Your job is to find out what the scientific evidence shows, then to educate the trier of fact on those issues. If that happens to bolster the case of the attorney who hired you, that is wonderful. But if not, then so be it. Your responsibility is to scientific truth. Now, I do recommend that if you see the evidence does not support the attorney's position, that you inform him or her as early as possible of the weaknesses in their case. But do not exaggerate or "stretch" to try and make their case. That is just not your role.

TYPES OF INVESTIGATIONS

The way in which you approach the examination and the level of detail given in your report can vary slightly with the type of investigation. No matter what the nature of the investigation is, you should certainly work for the highest quality of evidence possible.

Criminal Investigations

Conducting a criminal investigation involves adhering to specific legal requirements. These legal requirements should also form the basis for ethical rules. The most obvious legal requirement in the United States is the Fourth Amendment to the U.S. Constitution. The Fourth Amendment states

> The right of the people to be secure in their persons, houses, papers, and effects, against unreasonable searches and seizures, shall not be violated, and no Warrants shall issue, but upon probable cause, supported by Oath or affirmation, and particularly describing the place to be searched, and the persons or things to be seized.

This seemingly simple statement is the basis for a great deal of case law regarding appropriate searches and the seizure of evidence. One question that has been the center of a great many court decisions is what constitutes an unreasonable search or seizure. U.S. courts have consistently held that for a search to be reasonable, either there must be a warrant issued by a court, or there must be some overriding circumstances to justify searching without a warrant.

Civil Investigations

Civil investigation centers on some sort of civil litigation. A lawsuit involves the allegation of wrongdoing where the potential damages are financial rather than criminal. Civil litigation includes wrongful termination lawsuits, intellectual property infringement lawsuits, and similar litigation. These cases will sometimes involve forensic examination of evidence.

For example, in the case of a sexual harassment lawsuit, it may be required to forensically extract and analyze e-mails. The process will be very much the same as it is with criminal investigations. The exact same issues with chain of custody and with documenting your process are involved. One major difference is that in a civil investigation, it is usually the case that a company owns the computer equipment and can provide permission to extract and analyze evidence; thus, warrants are unlikely to be needed. It is important to know that a civil litigation can lead to criminal charges. Therefore, you should apply the same rigor as you do with a criminal investigation.

PRIVATE INVESTIGATOR LICENSES

Many states in the U.S. have laws requiring that anyone who is not a member of law enforcement hold a valid private investigator license in order to perform forensics. These laws have been controversial, and the American Bar Association has denounced them. However, they are still the law in many states. You should check laws in your state. If a private investigator license is required in your state, or if you intend to perform forensics regarding a case in a state that requires such a license, it is critical that you have one.

THE SCIENTIFIC METHOD

We concluded our discussion of ethics by stating your responsibility is to scientific truth. That necessitates an understanding of the scientific method. Unfortunately, many people, even those with extensive formal education, are unclear on the scientific method. One always begins with a hypothesis. Contrary to popular misconception, a hypothesis is not a guess. It is a question that is testable. If a question cannot be tested, then it has no place in science whatsoever. Once one has tested a hypothesis, one has a fact. For example, if I suspect that confidential documents were on your Android phone and subsequently emailed to a third party and deleted (my hypothesis), I can conduct a forensic examination of your computer (my test). If that examination finds evidence of deleted documents and an email to that third party, I now have a fact.

The next step is to build a theory of the crime, based on multiple facts. The fact that confidential documents were on your phone, while very interesting evidence, is not in and of itself enough. There are other explanations for how that could have been on your phone. So, we seek additional evidence. Did your phone have security such as biometric locks? That would rule out someone else using your phone. We can check phone GPS records to determine if you were in the location the confidential documents were taken from at the time they were taken.

Once you have collected enough facts, there must be some explanation of the facts. In science, that explanation is called a theory.[3] To be a good theory, it must account for all the available facts. For example, my theory of the crime might be that you stole the confidential documents with the intent to sell them to a competitor. I would need to check all the facts and see if they all match my theory. This process of forming a hypothesis, testing the hypothesis, and synthesizing facts into a cogent theory is the scientific method.

The Philosophy of Science

The scientific method is the cornerstone of all science. And forensics is a science, or at least should be. A forensic examiner should also be familiar with the philosophy of science. The philosophy of science is based upon two principles: verification and falsifiability. The idea of verification is simple. Test your hypothesis. Whenever possible repeat the test with a different tool. This will either verify or invalidate your evidence.

Falsifiability is a central tenant in science. Karl Popper was a philosopher who articulated the basic philosophy of science and advocated falsifiability as the center of science. Falsifiability means that it is possible to disprove something. The best way to understand this concept is to contemplate a counter example. Consider something that is not falsifiable. If I tell you that I did not place the incriminating files on my phone but someone else did. However, that someone else is invisible, does not emanate any heat, cannot be touched, felt, or sensed, and logs in with my username, there is absolutely no way to test that. It is therefore not falsifiable. Ideas that are not falsifiable have no place in science. They simply are claims that cannot be verified.

Peer Review

The Daubert standard discusses whether or not a technique or tool has been subjected to peer-reviewed publication. Peer review is another important issue in the scientific community. Peer review means that other professionals in that field have reviewed the work and found it to be valid. For example, if you wish to publish a paper on some forensic technique. If you submit that paper to a peer-reviewed journal, the first thing that happens is that the editor sends your paper (without your name on it) two or three experts in the field to review. These experts don't have to agree with your conclusions, but they must agree that the paper is valid, the processes appropriate, and the methodology viable. This occurs before the paper is even published.

The paper must describe what you did, with sufficient detail to allow anyone who wishes, to repeat your work and test your findings. Once your paper is published, any reader of that journal is free to test your findings and to write into the journal should they have issues with your conclusions or your methodology.

The concept of peer review is multiple layers of checks and balances. First by the reviewers that determine if the paper is even of sufficient quality for publication. Then by the readers, who tend to be professionals in that field. If an idea withstands such scrutiny, this is still no guarantee that it is completely accurate, but it is certainly a much high level of reliability than any other method we have. For this reason, peer-reviewed sources are usually considered more reliable than non-peer-reviewed sources.

Locard's Principle of Transference

Forensic science could be said to proceed entirely from Locard's principle of transference. Dr. Edmond Locard was a forensic scientist who formulated what has become known as Locard's exchange principle or Locard's principle of transference. This principle was first applied to physical forensics, and it essentially states that one cannot interact in any environment without leaving something behind.[4] For example someone cannot break into a house and not leave something. That something could be a fingerprint, a hair, a footprint, etc. Now a careful criminal will cover up some of this, for example using gloves to keep from leaving fingerprints. But something will be left behind.

This also applies to digital evidence. There are so many bits of data collected on a modern mobile phone. It is quite difficult for someone to perpetrate a crime without some remnant of evidence being left on their computing devices. This is why we conduct forensic analysis. Phone GPS records can contain traces of movement. Every text or call leaves a trace on the mobile device and in the cell carrier records. All of these are traces that need to be examined.

THE FORENSIC REPORT

At some point you will have to write a report of your forensic analysis and conclusions. This report will detail what tests you conducted and the results. If you took adequate and complete case notes, you should be able to use those as the basis for your forensic report. Your forensic report is essentially a summary of the case notes. The SANS Institute recommends three sections in your report[5]:

Overview/Case Summary: This is a general summary of what the case is about, and what the goal of the investigation is.

Forensic Acquisition and Exam: In this section you will discuss what you did. Starting with acquiring the forensic evidence, then your actual tests.

The United States Department of Justice[6] states

> The examiner is responsible for **completely** and accurately reporting his or her findings and the results of the analysis of the digital evidence examination. Documentation is an ongoing process throughout the examination. It is important to accurately record the steps taken during the digital evidence examination.

Research papers published as early as 2007[7] state that a digital forensics report should include list of a number of items that should go into a report including description of collection and examination procedures, all log files generated by forensic tools, description of steps taken during examination, etc.

The well-respected SANS Institute publishes guidelines[8] on digital forensics reports. They state a digital forensics report should include overview/case summary, details on the forensic acquisition and exam, details of all steps taken, etc.

The Reference Manual on Scientific Evidence: Third Edition[9] states "Under Federal Rule of Civil Procedure 26(a)(2)(B)(i), the expert report must contain the basis and reasons for all opinions expressed, and certainly the expectation is that oral testimony will do the same."(emphasis added).

You should note a commonality in the recommendations from these diverse sources. They all state that it is important to include a great deal of detail. A general guideline is that your report should contain enough detail that any competent forensic examiner could take your report and completely duplicate your tests and processes. This would allow that examiner to confirm or refute your findings. It is frankly quite difficult to have too much detail.

QUALITY CONTROL

We have addressed legal requirements for forensic examinations, testimony, and reports. We have also discussed ethical requirements of the forensic examiner. We have examined the scientific method, as well as recommendations on how to write a report. However, the fact is we are all human beings. Errors can occur. And as we discussed earlier in this chapter, a great deal is at stake in a forensic examination. It is important to minimize errors. That brings us to the issues of quality control.

Lab Quality

FBI RCFL

The U.S. Federal Bureau of Investigation Regional Forensics Laboratories is a good place to seek guidance on forensic quality. RCFL or Regional Computer Forensics Laboratory[10] is described by the FBI as: "An RCFL is a one stop, full-service forensics laboratory and training center devoted entirely to the examination of digital evidence in support of criminal investigations". This means a lab that is involved in forensics, and training. These labs offer a great deal of training options including continuing education courses[11] such as:

Case Agent Investigative Review
Image Scan Training
Social Media Evidence
Capturing a Running Computer System
Mobile Forensics"

Essentially the RCFL's are hubs for forensic information. That includes information on procedures and quality controls. These labs provide an excellent source for continuing education. They also provide a model for how a forensic lab should, ideally, be setup. If your laboratory meets the standards,[12] you can apply to enter into a Memorandum of Understanding with the FBI to become participating agencies in the RCFL.

American Society of Crime Laboratory Directors

Another source for information on laboratory quality is the ASCLD. The American Society of Crime Laboratory Directors (ASCLD)[13] provides guidelines for forensics labs of all types, not just digital forensics. The ASCLD also supports standards for certifying forensics labs. ASCLD offers voluntary accreditation to public and private crime laboratories in the United States and around the world. The ASCLD/LAB certification regulates how to organize and manage crime labs. Achieving ASCLD accreditation is a rigorous process and there are literally several hundred criteria that must be met. The process can often take more than two years to fully prepare for accreditation. It spends this time developing policies, procedures, document controls, analysis validations, and so on. Then, the lab needs another year to go through the process. However, the ASCLD does have courses to help prepare one for lab accreditation.[14] This course is described as

> *This course has been designed to assist crime laboratory personnel to prepare for ASCLD/LAB*–International. *ISO 17025:2005 and ASCLD/LAB*–International *Supplemental Requirements for Testing Labs will be reviewed, and preparation exercises will be provided. This program will serve to provide the attendees with a better understanding of the types of planning and activities which may be considered in order to prepare for the formal application and external assessment process. This program is designed to supplement internal management practices which a laboratory should have in place to support ASCLD/LAB*–International *preparation activities. Attendees familiar with the ASCLD/LAB Legacy program will be provided with opportunities to compare internal laboratory activities involved with each program. An attendance certificate will be mailed to the student.*[15]

It may be that the ASCLD process is cost prohibitive for your organization. However, referring to their standards to help develop your own guidelines for digital forensics standards in your lab is a good idea.

ISO 27037

The full name of this standard is "ISO/IEC 27037:2012 – Information technology – Security techniques – Guidelines for identification, collection, acquisition and preservation of digital evidence." This is an excellent reference when deciding what standards will govern your laboratory. The standard's abstract states "It provides guidance to individuals with respect to common situations encountered throughout the digital evidence handling process and assists organizations in their disciplinary procedures and in facilitating the exchange of potential digital evidence between jurisdictions."[16]

There are several related standards:

- ISO/IEC 27041 offers guidance on the assurance aspects of digital forensics e.g., ensuring that the appropriate methods and tools are used properly.
- ISO/IEC 27042 covers what happens after digital evidence has been collected i.e., its analysis and interpretation.
- ISO/IEC 27043 covers the broader incident investigation activities, within which forensics usually occur.

- ISO/IEC 27050 (in 4 parts) concerns electronic discovery which is pretty much what the other standards cover.

NIST

The United States National Institute of Standards has several standards relevant to digital forensics. You can begin perusing them at https://www.nist.gov/programs-projects/digital-forensics. Recall that when discussing specific mobile forensics tools, we mentioned the Computer Forensics Tool Testing (CFTT) standard. That can also be found via the NIST website.

Tool Quality Control

It is also important to ensure your tools are up to standard. In the last section we mentioned the CFTT standard. That is a good basis to begin selecting tools. There are two parts to tool quality control. The first is only selecting tools that have been widely accepted in the forensic community. Preferably where one can read peer-reviewed scientific studies on the tool. The second part is to actually conduct your own testing of the tool. There are two primary ways to test a given tool.

The first method is to conduct a test on a known item, with preset properties. In other words, if you are testing a tool for Android forensics, select an Android phone that you know what it contains. Either purchase a new phone, or completely factory reset an existing phone. They put specific data on it including calls, text messages, browser activity, deleting apps, etc. Then use the tool you are testing. You know exactly what the tool should find, so it is quite easy to determine the efficacy of the tool.

The second method is to test against a known good tool. If you are testing a new tool compare it to a tool you have used for some time. Use both tools on the same mobile device. It should be that the same data is retrieved by both tools. It is certainly possible that one tool has more features and might find a few items the other tool misses. That is not a quality concern. But any time both tools identify the same artifact, their findings about that artifact should be substantially the same.

Investigator Quality Control

Just as the laboratory must maintain quality, you have to maintain the quality of your individual analysts and investigators. Quality of forensics personnel includes ensuring they have the appropriate training and education. It also involves ensuring that they have a clean background.

Training & Education

It is important that forensic examiners have proper training. Ideally that means a computer-related degree along with specific digital forensics training. However, there are many digital forensic examiners without any background in computer science or related disciplines. In such cases the examiner should seek training to fill in the gaps in their technical background. Sometimes industry certifications can help with this. For example, CompTIA A+ certification provides basic understanding of computers, mobile devices, and networking. The CompTIA Network+ certification provides a working understanding of computer networks. Apple offers a number of certification courses https://training.apple.com/us/en/recognition.

The goal is for the examiner to have a thorough knowledge of forensic procedures, and to have a deep understanding of the underlying technology. Whether that knowledge is acquired via university education, professional education, or certifications is less important than the knowledge obtained. The examiner cannot simply blindly plug mobile devices into forensic tools and read out the output. The examiner must understand the technology involved.

Examination Quality Control

There are a number of ways to ensure the quality of examinations. Each of these can be implemented in a laboratory to help ensure quality control:

Rechecking: Performing a test with a second tool is an excellent way to validate the results of an examination. This is particularly useful for solo practitioners.

Spot checking: Having a colleague spot check some portion of an examination. This can be done periodically to validate findings.

Supervisor checking: This is like spot checking, only conducted by the supervisor.

Report review: Either a colleague or supervisor simply re-reads the examiner's report looking for gaps or issues.

In a laboratory all of these can be employed to ensure quality. It is also important to have a set frequency. For example, you might have 1 out of 10 examinations subjected to spot checking and 1 out of 20 subjected to supervisor checking. The specific periodicity is not as important as having a set frequency that you adhere to. Just as important, should you find an issue there must be a process to investigate and address that issue quickly.

CHAPTER SUMMARY

In this chapter we have discussed legal issues regarding forensic examinations and testimony. It is critical to be familiar with the requirements and rules in your jurisdiction. We also covered evidence tracking, ethics, and the scientific method. The final section of this chapter was about quality control. All of these topics are really about the same issue. That issue is ensuring a high degree of accuracy and integrity in forensic examinations. It is critical that every forensic examination be conducted in the most scientific manner possible and that the evidence produced be as reliable as possible.

CHAPTER 11 ASSESSMENT

1. What is the most affordable method of evidence tracking?
   a. Logs
   b. Software
   c. Barcode
   d. RFID

2. What U.S. Standard describes what is required of expert testimony in a federal court?
   a. Frye
   b. Rule 902
   c. Daubert
   d. Rule 903
3. What principle tells us that any interaction with a system is likely to leave some trace?
   a. Frye
   b. Daubert
   c. Falsifiability
   d. Locard
4. What is a scientific theory?
   a. An educated guess
   b. A testable guess
   c. A model explaining the evidence.
   d. A guess that has evidence
5. What is the general minimum standard for a forensic report?
   a. It must list the tools you used.
   b. It should be sufficient for someone to recreate your process.
   c. It should be sufficient to allow you to testify at court.
   d. It must list your qualifications.

NOTES

1 https://armypubs.army.mil/epubs/DR_pubs/DR_a/pdf/web/ARN13728_R195_5_FINAL.pdf
2 https://www.law.cornell.edu/rules/fre/rule_702
3 http://www.scientificamerican.com/article/just-a-theory-7-misused-science-words/
4 http://www.forensichandbook.com/locards-exchange-principle/
5 http://digital-forensics.sans.org/blog/2010/08/25/intro-report-writing-digital-forensics/
6 https://www.ncjrs.gov/pdffiles1/nij/199408.pdf
7 Abdalla, S., Hazem,S., Hashem, S. (2007). Guideline Model for Digital Forensic Investigation. Annual ADFSL Conference on Digital Forensics Security and the Law https://commons.erau.edu/cgi/viewcontent.cgi?article=1029&context=adfsl
8 https://www.sans.org/blog/intro-to-report-writing-for-digital-forensics/
9 National Research Council. (2011). *Reference manual on scientific evidence*. National Academies Press.
10 http://www.rcfl.gov/
11 http://www.rcfl.gov/DSP_T_CoursesLE.cfm
12 http://www.rcfl.gov/Downloads/Documents/Benefits_of_Participation.pdf
13 http://www.evidencemagazine.com/index.php?option=com_content&task=view&id=1159&Itemid=217
14 http://www.ascld-lab.org/training/
15 http://www.ascld-lab.org/preparation-course-for-testing-labs/
16 https://www.iso.org/standard/44381.html

Appendix A: Answers to Review Questions

CHAPTER 1

1. Which of the following has the longest wavelength?
   a. Radio waves
   b. Gamma Rays
   c. Ultraviolet
   d. X-Rays

2. _____ works by using an allocated band of frequencies and changing between the frequencies, using one frequency at a time.
   a. FHSS
   b. DSSS
   c. CSS
   d. THSS

3. The term _____ means there are multiple separate radio channels side by side within a designated radio band.
   a. Multiplexing
   b. Duplexing
   c. Modulation
   d. Orthogonal

4. _____ uses the entire bandwidth to broadcast a signal, and relies on a sinusoidal signal of frequency increase or decrease.
   a. FHSS
   b. DSSS
   c. CSS
   d. THSS

5. _________requires the use of guard bands.
   a. FDMA
   b. TDMA
   c. CDMA
   d. QDMA

6. When using ______ orthogonal codes are used to separate different transmissions. Each symbol to be transmitted is encoded using a specific code.
   a. FDMA
   b. TDMA
   c. CDMA
   d. QDMA

7. Pico cells cover _____.
   a. Home or office
   b. Small urban zones
   c. Mid-size area
   d. A building or tunnel

8. _____ is the part of the cell network responsible for communications between the mobile phone and the network switching system.
   a. BTS
   b. MSC
   c. VLR
   d. BSS

CHAPTER 2

1. What best describes an isometric antenna?
   a. An omnidirectional antenna
   b. A directional antenna
   c. An SDR antenna
   d. A theoretical ideal antenna

2. What is a concern with near field regions of antenna that is not a major concern with the far field region?
   a. Absorption of the radiation
   b. Distance attenuation
   c. Free-space path loss
   d. Power attenuation

3. What does the following equation describe?
   $\frac{P_r}{Pt} = \left(\frac{A_r A_t}{d^2 \lambda^2}\right)$
   a. Free-space path loss
   b. Friis equation
   c. Discrete Cosine Transform
   d. Fast Fourier Transform

4. Which of the following are voltages used with SIM cards?
   a. 1
   b. 1.8
   c. 3
   d. 3.8
   e. 5
   f. 5.8
   g. 7

5. The second two digits in an ICCID represent what?
   a. Country code
   b. Network identifier
   c. SIM number
   d. Issuer identifier

6. Where should you look to find the mobile network code (MNC)?
   a. First two digits of the ICCID
   b. Digits 4 – 6 of the IMSI
   c. First two digits of the IMSI
   d. Digits 4 – 6 of the ICCID

CHAPTER 3

1. What file system does the iPhone use?
   a. HFS
   b. HFS+
   c. EXT
   d. APFS

2. What kernel is used in iOS?
   a. Micro kernel
   b. Monolithic
   c. XNU
   d. Darwin

3. _____ causes the various system components to be loaded to different memory addresses each time the system is booted.
   a. Micro kernel
   b. ASLR
   c. APFS
   d. Address space layer randomization

4. Where is the core motion framework found?
   a. Core OS
   b. Touch layer
   c. Kernel
   d. Core services

5. Which version of iOS introduced 6 digit passcodes?
   a. iOS 9
   b. iOS 10
   c. iOS 11
   d. iOS 12

6. What is the primary signal used in Apple Face ID?
   a. Visible light
   b. Ultraviolet light
   c. Infrared light
   d. Radio waves

CHAPTER 4

1. _______ is the default flash file system for the AOSP (Android Open Source Project).
   a. APFS
   b. YAFSS
   c. F2FS
   d. JFFS2
2. Which directory should you look in for frequently accessed data?
   a. bin
   b. data
   c. cache
   d. sbin
3. What ADB command will list all the Linux commands supported by the device?
   a. adb shell Linux
   b. adb shell Command
   c. adb shell ls system/bin
   d. adb shell ls commands
4. What does the command *pm list packages -e* do?
   a. List only enabled packages
   b. List only executable packages
   c. List only excluded packages
   d. List only extraneous packages
5. What command will tell you if the device is OEM unlocked?
   a. adb shell oem device-info
   b. fastboot oem device-info
   c. adb shell device-info
   d. fastboot device-info
6. What is the Initial Program Load responsible for?
   a. loading system apps
   b. loading the first user apps
   c. setting up the external RAM
   d. setting up ART

CHAPTER 5

1. How many communication apps can most forensics tools acquire data from?
   a. Less than six
   b. All apps
   c. A few
   d. Most
2. According to SWGDE, what is the third level of forensic examination?
   a. Manual
   b. Logical
   c. File System
   d. Physical
3. A method that requires disassembling the phone and accessing the circuit board.
   a. JTAG
   b. Chip-Off
   c. MocroRead
   d. Physical
4. What is the best way to describe what must be in an expert report?
   a. The tools used
   b. The techniques used
   c. The process used
   d. The basis and reasons for all opinion

CHAPTER 6

1. The _____ is a unique address that identifies the access point/router that creates the wireless network.
   a. MAC Address
   b. SSID
   c. BSSID
   d. WAP Name
2. According to SWGDE what is the 2nd tier in the forensics pyramid?
   a. Manual
   b. Logical
   c. Physical
   d. JTAG
3. What is the best way to describe what must be in an expert report?
   a. The tools used
   b. The techniques used
   c. The process used
   d. The basis and reasons for all opinion

4. How many communication apps can most forensics tools acquire data from?
   a. Less than six
   b. All apps
   c. A few
   d. Most
5. What, if anything, is incorrect about this statement: adb backup -all -f?
   a. Nothing
   b. No file path
   c. No source path
   d. Should be from shell

CHAPTER 7

1. An ampere is a unit of what?
   a. Charge
   b. Current
   c. Resistance
   d. Power
2. A two terminal component that conducts current in one direction is called a _____.
   a. diode
   b. discrete device
   c. package
   d. printed circuit board
3. IEEE 1149.1 defines how many TAPs for boundary scanning?
   a. 3
   b. 4
   c. 5
   d. 6
4. (True/False) Chip-off is not as destructive as JTAG. **False**
5. (True/False) TRST is optional. **True**

CHAPTER 8

1. Records in a relational database are identified by ______.
   a. Record number
   b. Foreign key
   c. Record key
   d. Primary key

2. The statement SELECT * FROM Employees WHERE City='Plano' and LastName!='Smith' does what?
   a. Retrieves all records in the city of Plano with a last name of Smith
   b. Retrieves all records in the city of Plano or with a last name of Smith
   c. Retrieves all records in the city of Plano and not the last name of smith
   d. Retrieves all records that are not in the city of Plano or not the last name of smith

3. By default, the ORDER BY clause sorts in what order?
   a. Ascending
   b. Descending
   c. There is no default you must set it
   d. It varies

4. SQL Lite stores decimal numbers as what type?
   a. Float
   b. Double
   c. Real
   d. Decimal

CHAPTER 9

1. According to NIST what is the maximum range of a cell tower?
   a. 35 km
   b. 35 miles
   c. to 10 km
   d. to 10 miles

2. What is the FCC position on GPS data?
   a. Unlike CSA, GPS is precise
   b. GPS is usually inaccurate.
   c. GPS is more accurate than CSA.
   d. GPS and CSA are about equally accurate.

3. The following formula is used for what purpose?

$$E = \sum_{i=1}^{n} Di + /- \sum_{i=1}^{n} M_i$$

a. Determining the margin of error for Cell Site analysis.
b. Determining the range of a cell tower.
c. Determining the area serviced by a cell tower.
d. Determining position of a cell tower.

4. What operating system does Samsung use for Smart TVs?
   a. webOS and Android
   b. webOS and Tizen
   c. Android TV and Oray
   d. Tizen and Oray
5. Why use IP connection for ADB on a Smart TV?
   a. You should only use IP if you don't have a USB cable.
   b. You usually cannot connect via USB cable.
   c. It is faster.
   d. It is more reliable.

CHAPTER 10

1. The most common method of steganography is what?
   a. Symmetric
   b. Asymmetric
   c. LSB
   d. Hashing
2. What is the term for the file you hide data in with steganography?
   a. Payload
   b. Carrier
   c. Channel
   d. Cover
3. What is a key schedule?
   a. An algorithm for generating round keys from the cipher key.
   b. A schedule for changing your symmetric key.
   c. A schedule for changing your asymmetric key.
   d. An algorithm for storing symmetric keys.
4. What is the most widely used asymmetric cryptographic algorithm today?
   a. DES
   b. AES
   c. DH
   d. RSA
5. Which of the following is the best description for a rainbow table?
   a. A table of keys to attempt brute force on a symmetric cipher.
   b. A table of keys to attempt brute force on an asymmetric cipher.
   c. A table of precomputed hashes.
   d. A table of known passwords for password cracking.

CHAPTER 11

1. What is the most affordable method of evidence tracking?
   a. Logs
   b. Software
   c. Barcode
   d. RFID
2. What U.S. Standard describes what is required of expert testimony in a federal court?
   a. Frye
   b. Rule 902
   c. Daubert
   d. Rule 903
3. What principle tells us that any interaction with a system is likely to leave some trace?
   a. Frye
   b. Daubert
   c. Falsifiability
   d. Locard
4. What is a scientific theory?
   a. An educated guess
   b. A testable guess
   c. A model explaining the evidence
   d. A guess that has evidence
5. What is the general minimum standard for a forensic report?
   a. It must list the tools you used
   b. It should be sufficient for someone to recreate your process
   c. It should be sufficient to allow you to testify at court
   d. It must list your qualifications

Index

9780367633004